The Antarctic Politics of Brazil

Ignacio Javier Cardone

The Antarctic Politics of Brazil

Where the Tropic meets the Pole

palgrave
macmillan

Ignacio Javier Cardone
University of Sao Paulo
São Paulo, Brazil

ISBN 978-3-030-80160-1 ISBN 978-3-030-80161-8 (eBook)
https://doi.org/10.1007/978-3-030-80161-8

This Palgrave Macmillan imprint is published by the registered company Springer Nature Switzerland AG
The registered company address is: Gewerbestrasse 11, 6330 Cham, Switzerland

To Renata, my unconditional partner and the light of my life

Foreword

In terms of both its population and its size, Brazil is by some distance the largest country in South America. It is a leading member of the Mercosur trade bloc and plays a central role in the politics of the continent. Despite speaking Portuguese rather than Spanish, Brazil exerts a significant cultural influence, both regionally and globally, and has done much to shape wider perceptions of Latin America. In relation to the international politics of Antarctica, however, Brazil has long been overshadowed by the activities and interests of Argentina and Chile, the two most southerly countries in South America. Argentina and Chile both claim large sectors of Antarctica as part of their national territory, and both are founder members of the 1959 Antarctic Treaty, which continues to function as the international governing structure for the Antarctic continent.

While Brazil cannot compete with either Argentina or Chile in terms of the duration and importance of its Antarctic connections, it still has a fascinating history of engagement with the continent. In *The Antarctic Politics of Brazil: where the tropic meets the pole*, Ignacio Cardone sets out the history of Brazil's interests in Antarctica from the mid-twentieth century to the present in a way that is both accessible and engaging. It presents Brazil's Antarctic history on its own terms, but also recognizes the importance of the wider international context in shaping the country's interests in the far south. This is a study that is both national and international in its character, and in presenting a history of Brazil's involvement with Antarctica, this book also offers a unique perspective on that continent's recent history.

While recognizing that geopolitics have certainly played a key role in Brazil's Antarctic history, an important contribution of this book is to go beyond an exclusive focus on strategic thinking in explaining Brazilian engagement with Antarctica. Cardone shows how other factors such as environmental connections, economic development, regional politics and international scientific reputation have also been important in shaping Brazil's Antarctic politics. This approach allows for a move away from a narrow focus on Antarctica as a small subfield of Brazilian military history to open it up to a much wider range of themes and approaches. Once opened up in this way, there are tremendous opportunities for thinking differently not only about the history of Brazil and Antarctica, but also about the history of Antarctica and even perhaps about the history of Brazil.

Why is a tropical country interested in the southern polar regions? The subtitle of the book—'where the tropic meets the pole'—sets up a fascinating intersection of two geographical extremes. From a national perspective, this is a complicated relationship that connects to wider understandings of geography and identity, in the past, the present and future. For Brazilian Antarctic policy makers and practitioners, a better question, perhaps, is why shouldn't a tropical country be interested in the southern polar regions? From a global perspective, Brazil's interest in Antarctica reminds us of the growing awareness of the interconnectedness of the global environment that has taken place since the middle of the twentieth century. It is interesting to note from Cardone's research that in the idea of 'frontage' this sense of environmental connection has been present from the earliest days of Brazil's Antarctic interests. Today, of course, there is an even greater awareness that, in common with the rest of the world, melting Antarctic ice has the potential to raise sea levels with devastating consequences for Brazilians living in coastal lowland regions of the country. The fact that Antarctica has come to matter to Brazil is a fact that matters far beyond Brazil.

In some ways, *The Antarctic Politics of Brazil* can be read as a work that decentres Antarctic history. As well as shifting attention to a country not traditionally associated with the Antarctic continent, this is a book that moves away from the traditional focus on exploration and discovery to a focus on government policy, diplomatic negotiation and naval logistics. In the South American context, the decentring of Antarctic history involves looking beyond Argentina and Chile, to ask how other countries in the region have engaged with Antarctica. In comparison with its southern neighbours, there is less of an Antarctic consciousness in Brazil (despite having a popular beer brand named 'Antarctica'), and this relative lack of interest has an impact on its national and international politics.

In other ways, however, *The Antarctic Politics of Brazil* might be seen as a necessary recentring of Antarctic history towards an account of the way the majority of people in the world experience Antarctica. In September 1983, Brazil became a consultative member of the Antarctic Treaty System in the same month as India; two years later, China attained consultative member status. In the space of just over two years, the promotion of these three countries to full membership of the Antarctic Treaty meant that this status was now something shared by the majority of the world's population, where before it had been a quite significant minority. As Cardone shows, this change occurred in the wider context of debates about the continent's economic mineral potential and in response to growing international pressure around 'the question of Antarctica' at the United Nations. By committing to becoming a full member, Brazil made a major contribution to strengthening the Antarctic Treaty and at a time when its existence was under quite significant threat.

Cardone has faced a number of challenges in writing this book, not least the COVID-19 pandemic, which has been particularly devastating in Brazil. As Cardone demonstrates in his analysis, Antarctic scholarship is shaped by national and international politics, and humanities and social science research almost invariably finds itself at the bottom of the pile. Funding is limited and archival sources are frequently difficult to access. *The Antarctic Politics of Brazil* is a tremendous contribution to Antarctic scholarship, and all the more so given the difficult circumstances in which it was written. Not only does this book offer a helpful insight into a previously little known dimension of Antarctic history and politics, but it also serves as a model for other studies of Antarctica that move away from traditional subjects and approaches.

Bristol, UK
April 2021

Adrian Howkins

Preface

Nosso país não pode perder o futuro majestuoso que antevemos para o século XXI: conquistar dois mundos totalmente contrastantes. Sobre a linha do equador, a Amazônia lendária e ardente. No fundo do mundo, a gelada Antártida.

(Henriques, 1984)

This book is about a connection. A connection between two regions imagined to be distant and contrasting, but which actually are physically, culturally and politically associated. It was born from the question about why Brazil got entangled in Antarctica. Being an Argentine national with a strong cultural Antarctic identity, I wondered how the tropical neighbouring country became engaged in the South polar region. In other words, I wanted to understand that encounter between the tropic and the pole.

That inquiry led me to research Brazil's Antarctic politics during my master's studies. During that process, I realized the lack of a constituted literature on the issue and the difficulties that anyone interested in the matter have to confront to find consistent and systematic data. The literature produced on Brazil has many times been coloured by either the Brazilian territorial approach based on geopolitical arguments, the official discourse of the Brazilian Ministry of Foreign Affairs or both. Furthermore, such literature is rather dispersed and hard to find, and very little has been produced outside Brazil.

Thus, when I was invited by Daniela Sampaio to contribute, along with herself and Adriana Abdenur, to a chapter for the excellent *Handbook on the Politics of Antarctica* (2017), edited by Klaus Dodds, Alan Hemmings and Peder Roberts, I considered this an opportunity to propose an alternative vision of Brazil's involvement in the continent. And so we did. However, the limits imposed by a chapter extension convinced me that more should be done in terms of producing a reference work that could cover the history of Brazil's engagement in Antarctica from a comprehensive and analytical perspective, and that such an effort would require a whole book.

Knowing the important challenges of such an effort, I invited Daniela Sampaio to join the journey, which she kindly accepted. Thanks to her partnership, the book took shape, and we started the process of submitting the proposal. To our delight, Palgrave Macmillan immediately showed interest, and after a few adjustments, they accepted the proposal. Later, circumstances made Daniela leave the enterprise, but I'm indebted to her for her insights and contributions in those early stages of the project.

The process of writing the book ended up much bumpier that I first thought. Access to documentation, literature and individuals proved to be really hard to obtain, sometimes impossible. As I advanced in the writing, I found more and more questions that needed further research and elaboration. Contradictions, information gaps and lack of sources are common in the literature, even in that produced in Brazil, and to date, no special collection, specialized library or catalogue is available. The absence of an organized Antarctic humanities and social sciences community and specialized centres seems to have contributed to the almost complete oblivion into which the issue has fallen within the country.

In addition, personal circumstances and conditions imposed by changes in the political background ended up delaying the whole process. Worsening the situation, the COVID-19 pandemic broke out. With no prospects to access the diverse number of archival repositories due to lockdown and travel restrictions, I had to adopt a different strategy. Thanks to the amazing job of digitalization done by the *Fundação Getúlio Vargas*, o *Arquivo Nacional*, the *Biblioteca Nacional* and the *Commissão Interministerial para os Recursos do Mar*, that effort paid off.

While the archival research cancelled as consequence of the pandemic is still wanting, the book is a thorough, empirically based, systematic and honest analysis of the trajectory of Brazil's involvement in Antarctica from a political perspective. With no funding or financial support of any sort, the book I present here has required a number of personal sacrifices and effort. I'm confident that it will become a reference book on the issue and will give birth to further works and research on the matter.

To conclude, I would like to thank the numerous people who contributed to this project, either directly or indirectly. Although the opinion and analysis in the book are entirely mine and any error or omission is entirely my responsibility, this work would have not been the same without their contributions. Even though it would not be possible to single out every one of them, I want to thank the International Relations Research Centre of the University of São Paulo and its director, Professor Rafael Villa, for all their support, critical feedback and friendship. Also, I want to mention my gratitude to Mr. Wellington de Amaral Jr, who made possible my first visit to Antarctica as part of the Brazilian Antarctic programme and who became a dearest friend. Also, my appreciation to Professor Jefferson Simões, Professor Leonardo Faria de Mattos and Professor Andrés Zarankin, for the fruitful interchange of ideas, information, opinions and experiences. I'm also indebted with the CPDOC of the *Fundação Getúlio Vargas*, the *Archivo Nacional*, the *Biblioteca Nacional* and the Brazilian Navy for their excellent work in keeping the historical memory of Brazil alive and granting me access to their digital repositories. Once more, I would like to single out the important contribution of Daniela Sampaio, who was my partner in the early stages of the book. Also, my greatest gratitude to Adrian Howkins for writing the Foreword of the book, Judy Tither for the proof-reading, Maísa Edwards for her kindness and help with some final details and to all the editorial team of Palgrave Macmillan who assisted me during the process. Finally, this book would not have been possible without the support of my friends and family and, most especially, the affection and understanding of my beloved Renata, my lovely Sofia and the little miracle that came while I was writing, Matías.

São Paulo, Brazil — Ignacio Javier Cardone

Contents

1 **By Way of Introduction: Brazilian Geopolitical Thinking and Antarctica** 1
1.1 The Brazilian Geopolitical Thought: Origins, Evolution and the Role of Antarctica 1
1.2 Conclusions: Antarctica at the Margins of Brazilian Geopolitical Concerns 17
References 22

2 **Prelude to a Brazilian Antarctic Policy** 27
2.1 Towards New Horizons 27
2.2 The First Brazilian Initiatives 47
2.3 Brazil's Step into Antarctic Politics 58
2.4 Conclusions 68
References 69

3 **When the Tropic Met the Pole** 77
3.1 The Search for Consultative Status 77
3.2 And Thus, the Tropic Met the Pole 95
3.3 Brazil's Definitive Hold in Antarctica 109
3.4 Conclusions 127
References 129

4 **The Rise of the Newcomers** 137
4.1 Brazil's Political Duality 137
4.2 Mining, Environment and Development 145

4.3 *Under the Steering of Science* 160
4.4 *Conclusions* 169
References 171

5 Science and Environment at the Central Stage 177
5.1 *The Restructure of Brazil's Antarctic Programme: From the Research Networks to the 4th International Polar Year* 177
5.2 *The Consolidation of the Scientific Profile of PROANTAR: The Golden years* 192
5.3 *Rebuilding from the Ashes: The Tragedy of the OPERANTAR XXX and the Contradictions of the Brazilian Antarctic Programme* 204
5.4 *Conclusions* 220
References 222

6 Conclusions 229
6.1 *Brazil's Trends in Its Antarctic Involvement* 229
6.2 *Looking Ahead* 238
References 241

Index 243

Acronyms

AT	Antarctic Treaty
ATCM	Antarctic Treaty Consultative Meeting
ATS	Antarctic Treaty System
BIOMASS	Biological Investigations of Marine Antarctic Systems and Stocks
CCAMLR	Convention on the Conservation of Antarctic Marine Living Resources
CIRM	Inter-Ministerial Commission for Maritime Resources (Brazil, *Comissão Interministerial para os Recursos do Mar*)
CNPq	National Research Council (Brazil, *Conselho Nacional de Desenvolvimento Científico e Tecnológico*)
COMNAP	Council of Managers of National Antarctic Programmes
CONANTAR	National Commission for Antarctic Issues (Brazil, *Comisão Nacional para Assuntos Antárticos*)
CONAPA	National Committee on Antarctic Research (Brazil, *Comitê Nacional de Pesquisas Antárticas*)
CRAMRA	Convention on the Regulation of Antarctic Mineral Resources Activities
DMAE	Sea, Antarctic and Outer Space Division of the Ministry of Foreign Affairs (Brazil, *Divisão do Mar, da Antártida e do Espaço*)
EACF	Antarctic Station Comandante Ferraz (Brazil, *Estação Antártica Comandante Ferraz*)
EMFA	Joint Chiefs of Staff (*Estado Maior das Forças Armadas*)
ESANTAR	Antarctic Activities Support Station (*Estação de Apoio Antártico*)

ESG	War College (*Escola Superior de Guerra*)
IANTAR	Brazilian Antarctic Institute (*Instituto Antartico Brasileiro*)
IAU	Uruguayan Antarctic Institute (*Instituto Antártico Uruguayo*)
IBEA	Brazilian Institute for Antarctic Studies (*Instituto Brasileiro de Estudos Antárticos*)
IGY	International Geophysical Year
IPY	International Polar Year.
NSC	National Security Council (Brazil, *Conselho de Segurança Nacional*)
OPERANTAR	Brazilian Antarctic Expedition
POLANTAR	Brazil's National Policy for Antarctica (*Política Nacional para Assuntos Antárticos*)
PROANTAR	Brazil's National Antarctic Programme (*Programa Antártico Brasileiro*)
RAPAL	Meeting of Administrators of Latin American Antarctic Programmes (*Reuniøn de Administradores de Programas Antárticos Latinoamericanos*)
SATCM	Special Antarctic Treaty Consultative Meeting
SCAR	Scientific Committee on Antarctic Research
UNCLOS	United Nations Convention on the Law of the Sea

List of Maps

Fig. 2.1	Map of the division of the Antarctic Peninsula proposed by Castro's frontage theory and claims advanced by Argentina, Chile and the UK (*Source* The Author, based on '*Atlas de Relações Internacionais*' [Carvalho & Castro, 1960, p. 159])	45
Fig. 3.1	Map of locations visited during the OPERANTAR I (*Notes* 1. The Argentine Jubany station was renamed Carlini in 2012. 2. The Faraday station complex was transferred to Ukraine in 1996 and renamed Vernadsky. *Source* The author)	106
Fig. 3.2	Map of locations studied for the installation of the Brazilian Antarctic Station (*Source* The author)	119
Fig. 5.1	Map showing Brazilian infrastructure in Antarctica (*Source* The author)	199

CHAPTER 1

By Way of Introduction: Brazilian Geopolitical Thinking and Antarctica

1.1 The Brazilian Geopolitical Thought: Origins, Evolution and the Role of Antarctica

Antarctica is a region usually related to geopolitics. The involvement of Brazil with the white continent has been frequently explained through a geopolitical approach, sometimes combined with other economic and strategic interests.[1] Although the book challenges a vision based exclusively on geopolitics to understand Brazil's involvement in Antarctica, it also acknowledges the key role that geopolitics played in promoting and shaping the relationship between the tropical country and the southern polar region. Even today, many ideas of the Brazilian tradition of geopolitical analysis continue to exert an important influence in the way in which Brazil approaches Antarctica.[2]

Geopolitical imagination has been key in informing state policy in Latin-America, especially during the 1960s and 1970s.[3] However, as Kacowicz recognized, it is extremely difficult to assess geopolitics as a

[1] See, for example: Abdenur and Neto (2014), Child (1988), Dodds (1997), Gandra (2009, 2013), and Mattos (2014, 2015).

[2] Whether that tradition could be considered as a distinctive geopolitical 'school' is discussed in the conclusion to this section.

[3] Child (1986, p. 12), Colacrai (2003, pp. 23–25), Costa (1992, pp. 225–226) and Kacowicz (2000, pp. 81–82).

I. J. Cardone, *The Antarctic Politics of Brazil*,
https://doi.org/10.1007/978-3-030-80161-8_1

causal mechanism.[4] The intention here is rather to identify the place that Antarctica held in the geopolitical writings developed in Brazil. To that aim, instead of picking up the geopolitical works that tackle the issue of the strategic importance of Antarctica, this section provides an overview of the evolution of this tradition of thought in Brazil, and the place that Antarctica held in those works. In this way, it will be easier to assess the relative importance that Antarctica played as a region within the Brazilian geopolitical thought as a whole.

As the aim here is not to provide a thorough account of the geopolitical developments in Brazil, but merely to illustrate the basis on which geopolitical ideas were built, readers *au fait* with the elaborations of the Brazilian geopolitical authors will find this section dispensable.[5] However, to those not familiar with the geopolitical imagination in Brazil, this section will provide a glimpse of the ideas that were developed through the twentieth century, linking the physical aspects of geography with the human factors and, particularly, with politics. In particular, this section seeks to provide a broader understanding of the place that Antarctica played on the whole territorial conception of power developed in that country.

To begin, it can be stated that, since its very beginning as a political unit, there was a deep territorial conscience of the Brazilian elites,[6] as was probably the case with all other Latin-American states.[7] However, in the case of Brazil, the physical characteristics of its geography seem to have favoured the formidable expansion of its territory—in the sense that the Andes range and the tropical forest of Amazonia have prevented other countries from expanding to the East. Additionally, the pioneering spirit of the *bandeirantes*[8] ended up building the idea of a dominion

[4] Kacowicz (2000, p. 95).

[5] Some interesting and more extensive accounts could be found in: Costa (1992), Karol (2013), and Miyamoto (1981, 1985, 1987) in Portuguese; and Becker (1986) and Hepple (1986) in English. Tambs (1970), presents a list and a brief account of the geopolitical bibliography previous to the 1970s.

[6] Kelly (1984, p. 439) and Miyamoto (1981, p. 78).

[7] Following Kacowicz, 'the concern for territory has been paramount for the international relations of the South American region in the nineteenth century and the first half of the twentieth century as well.' Kacowicz (2000, p. 84).

[8] The *bandeirantes* were colonists who, from the sixteenth century on, penetrated the interior of what is actual Brazil, starting from São Paulo de Piratininga (São Paulo) and extending the Brazilian territory to the West.

of the will over the nature, establishing one of the core symbols of the constituting idea of the 'Brazilian.' That territorial conscience of Brazilian political elites led many authors to affirm a geopolitical orientation from early in the history of Brazil. However, that orientation or conscience did not constitute a distinctive body of knowledge.[9] Only during the third decade of the twentieth century did systematic elaborations develop what could be called a geopolitical tradition.

Even geography as a discipline would not mature in Brazil until the 1930s, falling well behind the advancements in Europe and the US. Only during the 1920s, modern geography was formed in Brazil with the arrival of Carlos Delgado de Carvalho—a Brazilian national born in France—and the writings of Everaldo Backheuser—a Brazilian geologist and engineer. Then, the inherited colonial geography, based on a nomenclature and memory effort, gave way to a more modern discipline, based on an analytical approach to the physical and human features of territory.[10] Nevertheless, that was not a straightforward and unresisted process.

The reform of geography was part of the 'positivist' spirit that was dominant for some parts of the Brazilian elites, which looked to modernize society through education and science.[11] That modernization was confronted not only with traditional values, but also with political and economic interests established through the political system, characterized by a strong federalism instituted during the so-called *Republica Velha* (1889–1930), and the consequent political phenomenon of *coronelismo*.[12]

[9] It could be stated that it constituted a tradition, particularly with respect to the way in which Foreign Policy was conducted. However, that tradition was not translated on a body of knowledge on the strict sense, that is as a systematic and explicit formulation of principles for interpreting reality and assisting decision-making. The tradition of Foreign Policy expressed itself as a few general principles that supposed to guide the action of Brazilian diplomacy. As principles of valuative nature, they eventually collided with some of the interpretations aroused from the geopolitical thought.

[10] In this respect, see: Anselmo and Bray (2002, pp. 113–117) and 'Apontamentos Bio-Bibliográficos - Prof. Carlos Delgado de Carvalho' (1944).

[11] Anselmo and Bray (2002, p. 112) and Foresta (1992, pp. 128–129).

[12] Coronelismo refers to a bargain system in which power relations developed during the First Republic in Brazil (1889–1930). After the establishment of a federalist system, state governments were given some autonomy, which was then reinforced by local oligarchies, represented by the '*coroneis*' or landlords. The *coroneis* guaranteed large numbers of votes in exchange for positions in the local government, which were provided by the states. By their part, the state governments would offer the votes from *coroneis*'

Both socio-political structures were underpinned by the prevalent monocultural exportation economic model and its consequent dependence on the international markets. Therefore, modernization opposed both the prevalent socio-political structures and the economic model in which it was based.

Despite the obstacles placed by the establishment, Delgado de Carvalho proposed a new approach to the Brazilian territory. That approach overcame the formalist interpretation based on the political administrative division, proposing instead the analysis of the physical and demographical features of the territory.[13] On those bases, Delgado de Carvalho and others, like Backheuser, started to propose changes not only in geography, but also in sociology, history and the educational system as a whole.[14] The constitution of a new and modern Brazil was the main objective of such changes, and those thinkers configured the vanguard that provided the intellectual background for the changes that followed.

In any case, that group was not free from discrepancies and schisms, and even disagreement regarding the secular or religious nature of education divided the efforts of educational reform.[15] Still, the shared, general view of that part of the Brazilian *intelligenzia* was the aggrandizement of Brazil as a nation, overcoming the traditionalism of Brazilian society to become a 'modern' nation, driven by science and education. In their view, what was at stake was, so to speak, the taming of the tropical nature of Brazil and its inclusion in the modern era of industrial development. Therefore, the tropical identity, the will to control and dominate nature, and the drive on the part of the elites to modernize shaped Brazil's geopolitical thoughts and produced its first systematic elaborations.

domains to the presidency in exchange for their political support. For more, see: Carvalho (1997) and Leal (1948).

[13] See: Castro (2009, p. 29), Costa (2009, pp. 41–42), Ferraz (2009, pp. 94–97), Penha (2009, pp. 118–119), and Sampaio Machado (2009, pp. 55–59).

[14] Those efforts ended up hatching in the 1920s in institutions such as the *Academia Brasileira de Ciências* (Brazilian Academy of Sciences) and the *Associação Brasileira da Educação* (Brazilian Association for Education) (Anselmo & Bray, 2002, p. 110).

[15] Anselmo and Bray (2002, pp. 114–117).

If Delgado de Carvalho is usually acknowledged as the founding father of modern geography in Brazil, Backheuser is recognized by the literature as being the father of Brazilian geopolitics.[16] Whether Carvalho contributed by adding the analysis of the human aspects involved to the physical characteristics of the territory, it was Backheuser who developed a characteristic 'geopolitical' approach. Backheuser's geopolitical scheme was not innovative in the sense that his proposal did not advance new ideas or theoretical explanations but presented an adaptation of the work of European geopoliticians within the Brazilian context.

By 1924, in the context of the interwar period and the accumulated exhaustion of the primary exportation economic system in Brazil, Backheuser wrote his first geopolitical work, built on many of the ideas already developed by the German school of geopolitics. Although he allegedly based his work on the ideas of Kjellén and recognized the importance of the innovating ideas of Ratzel,[17] he was actually inspired in his writing by many of the same ideas offered by Haushofer and his group.[18] In those writings, Backheuser proposed a scheme in which the territory was thought of through the political implications to the nation. Repeating the organicist evolutionary analogy of the German school,[19] Backheuser thought of the State as a living organism fighting for its survival on a hostile environment. In that environment, the physical features of the territory and the human factors played a determinant role. For that author, elements such as the shape of the territory, its borders, the territorial division, the distribution of population and resources, the 'race' of the population and the geographical features were definitive for the analysis of the threats and the potential for development.

In the case of Brazil, those elements were translated into the expansion to the West, the relocation of the capital city in the hinterland, the interconnection through modern and development communication networks, the dynamism of the border communities, the education of the elites and the general public, and the 'whitening' of the population. These concerns were oriented to create conditions that could overcome the tropical

16 Anselmo and Bray (2002, p. 110), Miyamoto (1981, p. 78), and Tambs (1970, p. 73).

17 Backheuser (1952).

18 Anselmo and Bray (2002, p. 110) and Costa (1992, p. 197).

19 Backheuser (1952).

limitations of Brazil, not only in its geography, but also in its human aspects. Relocating the capital and expanding a network of communication were pivotal to the expansion to the West, bringing development to the Amazonian wilderness, allowing the extraction of the abundant resources available and consolidating Brazil's territorial dominance. At the same time, the whitening of the population[20]—through miscegenation and European migration—was key to overcome the limitation of the 'laziness' of the indigenous and black people.[21]

The projection of power beyond borders was not proposed by Backheuser, who was more worried about the consolidation of Brazil's acquired territory. He regarded the Atlantic as a huge oceanic barrier and, thus, as a favourable factor. However, he did not consider it a mean to project Brazil's power beyond its coasts. The same view was shared by his contemporaries. That is probably why Antarctica is not mentioned in his work nor in that of his colleagues during those first decades of geopolitical elaborations.

Almost coeval with those writings, Mario Travassos, by that time Captain of the Brazilian Army, wrote one of the most influential geopolitical works in Brazil: *Aspectos Geográficos Sul-americanos* (South American Geographical Issues).[22] That work was published immediately after the 1930 world economic crisis that had a strong negative impact over the Brazilian monocultural primary economy. The impact of the crisis led the political order that sustained that model to its terminal crisis. A *coup de etat* by the so-called *Revolução do 1930* (1930 Revolution) replaced the government and installed a provisional Presidency under Getúlio Vargas' rule. After taking the government, Vargas suspended the 1890 Constitution and governed by decree, intervening in the government of almost all the Brazilian states.[23]

[20] Anselmo and Bray (2002, pp. 111–113) and Costa (1992, pp. 196–197).

[21] In Backheuser, this geographic and racist determinism is relatively strong. For that author the climate and the 'race' of Brazil population are contrary to the industry and development of the country, and it is only through the promotion of the 'whitening' of the population that the population could be transformed on an industrious people that would overcome the climatic and geographic barriers. However, the idea of the 'whitening' of the population as necessary for national development was a diffused idea on the Brazil of that time. Anselmo and Bray (2002, p. 112).

[22] Later re-edited under the title: *Projeção Continental do Brasil* (Brazil's Continental Projection), and which is the most referenced work of that author.

[23] Landolfi and Grynszpan (1987).

Under Vargas, Brazil started a process of rapid industrialization. The crisis of the primary exportation economic model and the growing urbanization increased the opportunities for the development of urban industries of low specialization. Such a process was in line with the modernist positivist ideologies of some part of the Brazilian elites that supported the new government. Educational, territorial and political reforms proposed by authors such as Delgado de Carvalho, Everardo Backheuser and Mario Travassos found a receptive audience in the new government, willing to promote a more centralized political system that could replace the outdated regionalist structure of power derived from the *coronelismo*. It is necessary to bear in mind that, by the middle of the 1930s, nationalism was a growing force in almost all Latin-American countries, and the authoritarian fascist experience in Italy and Spain and the Nazi movement in Germany were regarded with admiration for their development achievements. The geopolitical thought seemed to provide the ideological basis over which many of the government authoritarian and centralized policies could be propelled, offering to those intellectuals the opportunity to promote many of the long-awaited reforms.

Travassos' elaborations directed their attention to similar issues to those worked by Backheuser but adopted a different perspective. Without an explicit acknowledgement of the influence of Halford Mackinder, Travassos extrapolated that author's scheme to the South American reality, identifying the 'heartland' of the subcontinent in the axis of Santa Cruz de La Sierra-Cochabamba-Sucre.[24] Considering it South America's geostrategic pivot, that area was regarded by Travassos as the key to regional hegemony. In his view, that heartland articulated the two main antagonisms in the region: the opposition between the (South) Pacific and the (South) Atlantic, and between the Amazon and Del Plata basins.[25]

Travassos saw in the South Atlantic the most dynamic of the two oceanic regions, due to its favourable position in world commerce, offering access to the markets of Europe and North America. Thus, the Amazon/Del Plata antagonism held the key for sub-continental hegemony. In that dynamic, the Santa Cruz de la Sierra-Cochabamba-Sucre axis was the geographic pivot through which the Pacific production could

[24] Foresta (1992, p. 129) and Miyamoto (1981, p. 79).

[25] Travassos (1947, pp. 41–43).

be exported to the Atlantic markets through one of the two antagonist basins.[26]

In Travassos' view, Argentina was in a better position thanks to its investment in infrastructure, being able to provide an exit for Bolivian products. Nonetheless, he considered that the political and natural features favoured Brazil, that it just needed to develop a similar policy in the Amazon basin. In contrast to the Del Plata, the Amazon basin was almost completely inside Brazil's territory, and the Brazilian ports offered a relatively advantageous position in relation to the northern markets. Consequently, the development of the hinterland was key to promote Brazil's hegemony. For the geopolitician, it was expected that, in the future, a sub-continental conflict could arise as a result of the opposition between the dominant continental antagonisms.[27]

Thus, the writings of Travassos presented important differences but also some similarities with the work of Backheuser. Travassos paid great attention to the communication system and its potentiality. Other issues considered important by Backheuser, such as the availability of natural resources and the population, are relatively absent in Travassos' work. Similar to Backheuser, the main concern of Travassos is directed to the development and the dominion of the interior of Brazil and not to an overseas projection of power. The Atlantic occupies a central place in Travassos' scheme as the most dynamic region of the world, presenting an opportunity to Brazil. However, that role of the Atlantic is not oriented to project their power overseas but to the interior of the continent.[28] As a consequence, Travassos did not pay any attention to Antarctica, which remained outside the scope of his works.

Brazil's new political momentum allowed the implementation of many ideas embedded in positivist and modernist ideologies. According to Bertha Becker,[29] geography as a discipline was institutionalized in the 1930s through the establishment of three complementary institutions:

[26] Travassos (1947, pp. 60–77).

[27] It is not surprising then that the position of Travassos generated anxieties not only in Argentina, but also in all other neighboring countries (Child, 1986, pp. 45–46).

[28] This is strongly emphasized on the title of the re-edited version of the book.

[29] Becker (1986, pp. 157–158). Also in: Anselmo and Bray (2002, p. 115) and Sampaio Machado (2009, p. 56). However, the latter identifies the creation of the *Curso Livre de Geografia Superior* in 1926 (under the impulse of Delgado de Carvalho and Backheuser) as the initial point of the institutionalization of geography in Brazil.

the creation of university courses in geography at São Paulo (University of São Paulo), in 1934, and Rio de Janeiro (Federal District University—currently Rio de Janeiro Federal University), in 1935; the establishment of the *Associação dos Geógrafos Brasileiros* (Association of Brazilian Geographers) in 1935[30]; and the establishment of the Brazilian Geographical Council in 1937, which was subordinated to the National Statistical Institute, created in 1934.[31]

Those institutional developments took place in the context of confrontation between liberal and catholic ideologies. The schism between the 'liberal spring'—exemplified on the figure of Delgado de Carvalho—and the 'catholic' spring—exemplified on the figure of Backheuser—expressed the division in the field of geography and education. In that confrontation, the influential role of the 'French school' of geography, through the active involvement of prominent geographers such as Vidal de la Blache, Pierre Deffontaines and Pierre Monbeig, ensured the triumph of the liberal position.[32] Such ideological division determined that geopolitical works continued to be developed, with only a few exceptions, not by geographers but by professionals from other disciplines.[33] That left geopolitics mainly outside of Brazilian academia and the universities. Thereafter, several other authors produced geopolitical

[30] Karol locates the creation of the AGB in 1934 (Karol, 2013, p. 52). The chronology followed here has been that of Becker (1986, p. 158).

[31] The National Statistical Institute was renamed *Instituto Brasileiro de Geografia e Estatística* (IBGE – Brazilian Institute of Geography and Statistic) in 1938.

In respect to this chronology, there are several discrepancies on the literature. The chronology presented here is based on the details present in the Brazilian Legislation: Decreto N° 24,609, 1934; Decreto N° 1,200, 1936; Decreto N° 1,527, 1937 and Decreto-Lei N° 218, 1938. Following the text of the decrees, a National Statistical Institute was instituted in 1934. The same decree summoned a Convention, which took place in 1936 and after which the National Statistical Council was established as a collegiate organ. The next year, the Brazilian Geographical Council was created as a dependency of the National Statistical Institute. Finally, in 1938, the National Statistical Institute was renamed Brazilian Institute of Geography and Statistics (*IBGE*) and the two collegiate organs—of Statistics and Geography—were named as National Councils (thus, the Brazilian Geographical Council being renamed as National Geographical Council).

[32] Becker (1986, pp. 158–159). See also: Anselmo and Bray (2002, pp. 113–117) and Hepple (1986, p. 80).

[33] Costa (1992, pp. 187–188).

works, but they did not attain the attention and influence that Backheuser and Travassos' work did.[34]

It would only be after the World War II, with the creation of the *Escola Superior de Guerra* (War College—ESG),that a new generation of geopoliticians aroused. The ESG, a higher military academy inspired on the US National War College and the French *Institut des Hautes Etudes de la Defense Nationale* (Institute of Higher Studies of National Defence),[35] was instituted in 1949.[36] It gave the new generation of geopoliticians an opportunity to produce their writings and disseminate their ideas within the military.[37]

Those works were framed within the context of the Cold War, with the confrontation between the Western and communist blocs. Therefore, the geopolitical analysis adopted a global perspective, identifying the challenges to Brazil in the broader context of continental defence and bipolar confrontation. That change in the scope of the geopolitical perspectives, and the international prominence that Antarctica acquired during the 1950s, explains why Antarctica appeared—even marginally—in the geopolitical considerations of the Brazilian authors in the late 1950s. Amongst authors, the most relevant and influential figures were the geographer Terezinha de Castro—who worked in closely partnership with Delgado de Carvalho—and the military generals Carlos de Meira Mattos and Golbery de Couto e Silva.

Golbery de Couto e Silva was the most influential of those three. Lieutenant Colonel by the time of his incorporation to the ESG,[38] Silva became involved on the military coup of 1964 and became a close advisor of Ernesto Geisel during his Presidency.[39] As a geopolitical author, Silva presented a well-developed model that represented an advance from previous elaborations, incorporating important geographic, demographic

[34] Miyamoto (1981, p. 79).

[35] Hepple (1986, p. 81).

[36] Created by Lei Federal N° 785, 1949.

[37] Costa (1992, pp. 214–216), Hepple (1986, pp. 81–83), Kelly (1984, pp. 444–445), and Miyamoto (1981, p. 80).

[38] Golberi de Couto e Silva Dicionário Histórico-Biográfico Brasileiro – DHBB. Source: http://www.fgv.br/cpdoc/acervo/dicionarios/verbete-biografico/silva-golberi-do-couto-e. Accessed on: 1 May 2018.

[39] Ernesto Geisel was President of Brazil between 1974 and 1979, under military rule. More on Geisel in Chapter 2.

and economic empirical data. Despite many of the topics addressed by Silva being already present in the previous generation of geopolitical works,[40] he presented a somewhat innovative perspective, based on what he called the 'theory of hemicycles': a series of circles radiating out from Brazil as centre in an azimuthal equidistant perspective determined the geographical distance—and thus the level of importance—from other regions of the world, especially in respect of the possibility of a military attack.

The perspective introduced by Silva was global in scope. It resulted from the international background of bipolar competition and the experience that Brazil's active participation in the World War II gave to his generation.[41] Brazil's involvement in the world conflict made many within the military and the political elites to start searching for a more relevant and active role in the international system. Silva's vision was permeated by the Cold War competition; thus, the emphasis was located on the possibility of communist aggression or internal subversion.[42]

In that regard, the author downplayed the possibility of external military aggression. His model of hemicycles determined that the inner hemicycle—the most critical due to the possibility of a direct and unexpected attack—presented no serious threat to Brazil due to the great distance from any potential centre of aggression. By contrast, the risk of internal subversion was a serious concern in Silva's writing. To counter that possibility, the emphasis was on development. For the author, the key was to adopt policies that could promote the welfare of the population and avoid any political and social discontent that could be used by the communists to promote internal unrest. Through that lens, the previous themes of territorial redistribution, the relocation of the capital, the development of the communication network and the advancement to the west reappeared in Silva.

In regard to Antarctica, the continent was located in Silva's inner and middle hemicycles, being vital for hemispheric defence as a 'superior

[40] Silva explicitly acknowledged the influence of authors such as Backheuser and Travassos in his work.

[41] Silva and Mattos were part of the Brazilian Expeditionary Force sent to fight in Italy at the end of the World War II.

[42] The analysis is based on Silva's *Geopolítica do Brasil* (Brazil's Geopolitics) (1967) which compiles several of his most influential and developed geopolitical writings on the 1950s.

manoeuvring turntable.'[43] However, the southern region was considered a relatively safe 'frozen desert,' untouched by political turmoil. That is why, in the relatively marginal discussions of Antarctica, the author emphasizes its importance, while at the same time described it as relatively safe.[44] When compared to the importance given to the African continent[45] and the possibility of internal subversion,[46] it is clear that Antarctica presented only marginal concern for the author.

In contrast, the geographer Terezinha de Castro—also instructor at the ESG—made Antarctica her gateway into geopolitical studies. Castro introduced the Antarctic continent into the geopolitical issues in Brazil and defended the importance of the frozen continent. Writing alone[47] and in collaboration with Delgado de Carvalho,[48] Castro developed the so-called *Teoria da Defrontação* (frontage theory),[49] which established the first authentic Brazilian political approach to the question of Antarctica.

For Castro, Antarctica was important for hemispherical defence but also because of the potential economic and strategic resources available there. Stressing the possibility of 'division' of Antarctica, Castro defended the active involvement of Brazil into the politics of the continent and called for the advancement of a territorial claim by Brazil. Her proposal was to divide the continent projecting the extreme coastal continental lines to the South Pole, allegedly following the 'sector theory' proposed by the Canadian lawyer Pascal Poirier for the Arctic. Notwithstanding her proposal included as part of the Brazilian sector a significative portion of the Argentine and Chilean claimed sectors in Antarctica, Castro appealed to a sub-continental brotherhood under the leadership of Brazil.

In agreement with Silva's view, Castro considered the South Atlantic as a critical region for hemispherical defence, favouring Brazil's overseas

[43] Silva (1967, pp. 84–85, 130–131).

[44] Ibid., pp. 80, 85–85, 136, 213.

[45] Ibid., pp. 91, 135–136, 237.

[46] Ibid., pp. 37, 46, 192, 215, 37, 46.

[47] Castro (1959, 1976).

[48] Carvalho and Castro (1956, 1960).

[49] As this issue is addressed in length in Chapter 2, the theory is not analysed in detail here. In this chapter, the attention is located on the relative importance of the issue within the author's work.

projection of power. This included the need to develop a foreign policy with greater autonomy from the US.[50] On the other side, opposing Silva's later view on the issue, Antarctica represented one of the main concerns of national defence, due to the possibility of communist aggression launched from that region, including the prospect of a 'climatic war.' Despite numerous inaccuracies and contradictions in her writings about Antarctica—such as overestimating in more than 2,000 km the distance between the different continents and the Antarctic—[51] Castro's proposition ended up being really influential in intellectual and governmental circles, being, until today, one of the main references of Brazil on what regard a Brazilian territorial position.

During the 1950s, Carlos de Meira Mattos became an instructor in several military academies. Like Silva, Mattos was involved in the military coup of 1964 and the military regime installed afterwards. In 1967, he took the ESG course and assumed the position of deputy of military affairs in that same institution.[52] His geopolitical writings started in 1958[53] but it was only in the 1970s that they became more influential.[54] Some analysts consider Mattos the most developed geopolitical author of his generation,[55] having presented a more systematized model of understanding of international politics through their geographical and human aspects. Ascribing to the realist school of International Affairs, Mattos envisaged Brazil as an emerging power, with the potential to become one of the major powers by the turn of twenty-first century.[56]

In terms of the continental projection of power, Mattos re-edited Travassos' formulation, identifying in the Amazon basin the strategic axis of South America.[57] Also framed by the context of the Cold War, Mattos

[50] Penha (2009, pp. 130–132).

[51] Castro (1959, p. 238). See Chapter 2 for more details.

[52] Carlos de Meira Matos. Dicionário Histórico-Biográfico Brasileiro – DHBB. Source: http://www.fgv.br/cpdoc/acervo/dicionarios/verbete-biografico/carlos-de-meira-matos. Accessed on: 1 March 2018.

[53] Tambs (1970).

[54] Hepple (1986, pp. 82–83) and Kelly (1984, pp. 441–442).

[55] See, in particular: Kelly (1984, pp. 441–442). Hepple (1986, p. 83) do not contradicts Kelly's argument, but considers Silva as notably more influential than Mattos.

[56] Mattos (1977, pp. 28–44, 78–102).

[57] Ibid., pp. 90–93.

considered the Soviet expansionism as the main threat to Brazil's national security, acknowledging the economic and social development as the key to secure Brazil's future and promote it as a world power.

In terms of the physical determinants, Mattos followed the 'morpho-political' analysis of Backheuser, considering the characteristics of Brazil as favourable. On the other hand, he adopted an opposite view to that of Backheuser regarding the population. In his view, the blending of races and the wilderness challenges they confronted installed the conditions upon which a new 'tropical civilization' destined to greatness would develop.[58]

In terms of the external projection of power, Mattos, like Castro, saw in the Atlantic the natural way to project Brazil's power overseas. Therefore, he emphasized the significant opportunities that the ocean offered[59] and the importance of counting with a dissuasive maritime strategic force. He argued on the essential weight that an agreement between the South American and the South African countries would have in establishing a South Atlantic defensive alliance, where a more active role in the Western world commerce defence could be assumed.

In relation to Antarctica, he considered the region a strategic target. Writing after Brazil's accession to the Antarctic Treaty (AT),[60] in the few pages he dedicated to the issue, Mattos defended the possibility of Brazil's participation in Antarctic scientific activities.[61] However, at the same time and following Castro's frontage theory, he supported the Antarctic territorial claims of Argentina, Brazil and Chile as defined by the so-called theory[62]—although without the importance that the issue had in Castro's works. In his view, the Brazilian claim in Antarctica did not conflict with any other occupation or activities already undertaken by other countries.[63] With Argentina, the US and the United Kingdom

[58] Kelly (1984, p. 451).

[59] Mattos (1977, p. 88).

[60] Brazil signed the Antarctic Treaty in 1975. This is addressed in detail in Chapter 2.

[61] See, for example: Mattos (1977, pp. 121–124).

[62] As it will be developed further in the book, Brazil actually never made an official claim to Antarctic territory, merely 'reserving' its rights in the future.

[63] Mattos (1977, p. 124).

already active in the coveted sector when Mattos wrote his book, he was certainly misinformed in this point.[64]

Silva and Mattos represented the peak of influence of the geopolitical thinking in the Brazilian government.[65] Some analysts consider them responsible for the promotion of many governmental policies of high impact, such as the Amazonian Pact of 1978 and Brazil's accession to the AT. Added to their role in the formation of a whole generation of the military intellectuality through the ESG, they exerted a direct governmental function that gave them access to the centre of political decision of the country.[66] This is also the reason why the ideas proposed by those authors generated anxieties in other South American countries, especially in Argentina.[67] The idea of a Brazilian 'continental expansion' and the possibility of its involvement in Antarctica were perceived as part of a Brazilian imperialism, if not merely an expression of US policy through one of its subordinated nations.

The influence of the ESG in the military elite established what Wanderley Messias da Costa called the 'geopolitics of the generals.'[68] That concept of power, which extended from the 1950s to the 1970s, reduced the national-territorial and security issues to the military orbit of the state. The situation changed with the advent of the works of the geographer Bertha Becker in the 1980s. Becker renovated the geopolitics field in Brazil, being the first author to criticize the previous works and complain about the lack of involvement of geographers and universities in the geopolitical debate. Having started as a tutor at the *Instituto Rio Branco* (the Brazilian diplomatic institute) in the 1960s, her publications acquired notoriety in the late 1970s and the 1980s, when she started to train a whole generation of geographers.

[64] At least, during the International Geophysical Year (IGY), it can be listed: the Laurie Island and Belgrano Argentine stations, the US Ellsworth station and the Shackleton, Signy Island and Base M (South Georgia) British stations. It is necessary to highlight that the Argentine Laurie Island station is the longest uninterrupted presence in Antarctica, which makes Mattos statement quite shocking.

[65] In this respect, see: Costa (1992), Foresta (1992), Hepple (1986), and Kelly (1984).

[66] This is something that distinguishes these two authors from the others. That is more evident in Silva, who was the main Presidential advisor during Geisel's Presidency (1974–1979), when Brazil acceded to the Antarctic Treaty.

[67] See, for example: Mastrorilli (1977) and Antártida Argentina: ¿Qué pretenden los brasileños? (1976).

[68] Costa (1992, p. 214).

Without signalling a complete break with the traditional geopolitical perspectives, Becker brought some new issues into her works, such as the environmental question. Centring her analysis on the Amazon, she focused on the regional development, the expansion of the resources frontier and the relations between the State, the territory and the international system in a more integrated world. Those considerations were influential in the elites and governmental circles and were adopted as principles for the design of several State policies.

Even when Eduardo Karol contends that Becker did not escape from the State dependency and the traditional approach of Brazilian geopolitics,[69] her contributions were significant. She left behind some of the previous perspectives centred on the national security and balance of power, and adopted a view more interested in economic and social development.[70] The general emphasis on economic integration and issues related to the new globalized economy led Kacowicz to name that new perspective as 'geo-economics.'[71] With her preoccupations elsewhere, Becker paid no attention to the Antarctic question.

In the twenty-first century, after Luiz Inácio 'Lula' da Silva assumed Brazil's Presidency (2003–2011), there has been a rising interest on geopolitics. Under 'Lula' da Silva, Brazil adopted a much more active role on the international stage, pursuing a higher profile that would justify its claim to be an emerging power. Giving renewed emphasis to the National Defence Policy, 'Lula' da Silva's policies led some academics to resume the analysis of 'traditional' geopolitical issues, recovering many of the writings of the abovementioned authors.

Brazil's status as emerging power corroborated Mattos' thesis and led to an updated interest in national security issues. Therefore, the recent geopolitical writing addressed issues such as: the Brazilian projection of power in the South Atlantic; the South-South cooperation, in particular with Africa; the need of a dissuasive military force to protect the national territory and the maritime space; and the active role to be played in the concert of nations.

[69] Karol (2013).

[70] In Becker, the emphasis on development is not fundamentally related to National Security—as a way to avoid internal subversion—as it was on Silva and Mattos.

[71] Kacowicz (2000, pp. 91–93, 97).

Regarding Antarctica, a handful of recent works affirmed the geopolitical importance of the continent.[72] The inclusion of Antarctica into the definition of the country's 'strategic surroundings' in the 2012 version of the *Política de Defesa Nacional* (National Defence Policy) and the mention to the South Atlantic region in the previous version was considered as revealing the strategic importance of the white continent for Brazil.[73] However, as Abdenur & Neto recognized,[74] the reference is merely occasional. Moreover, despite the discursive reference to the strategic importance of Antarctica for Brazil, some works recognized the difficulties that the programme experienced in obtaining the necessary funding.[75] In all, apart from some mention to the strategic importance of scientific research for ensuring Brazil's participation in the Antarctic regime, there is little originality in recent analysis about Brazil in Antarctica.

1.2 Conclusions: Antarctica at the Margins of Brazilian Geopolitical Concerns

Despite some authors attributing originality to Brazil's geopolitical thought, ascribing it the character of a distinctive 'school' of Geopolitics,[76] there are good arguments to agree with Miyamoto in that there is no fundamental innovation in Brazilian geopolitical works.[77] Brazilian geopolitical analysts did constitute an important 'tradition' and exerted a considerable influence in many governmental policies. However, there is no original theoretical substantial development but the use of ideas originated in Europe and the US, adapted to the Brazilian context. Whether they located Brazil at the centre of their concerns—both theoretically and geographically—they did not elaborate a distinctive approach to geopolitics nor to geography.[78]

[72] Abdenur and Neto (2014), Costa (2012), Gandra (2009, 2013), and Mattos (2014).

[73] Abdenur and Neto (2014) and Mattos (2014).

[74] Abdenur and Neto (2014, p. 10).

[75] Câmara et al. (2021), Cardone (2014, 2015), Gandra (2009, p. 72, 2013, p. 152), and Sampaio et al. (2017).

[76] Carmona Gomes (2012), Child (1988), Foresta (1992), and Tambs (1970).

[77] Miyamoto (1981). Costa (1992), presents a similar argument.

[78] Karol (2013).

Following the previous analysis of the geopolitical imagination in Brazil, it is possible to identify some trends on different periods, which are summarized in Table 1.1. The development of geopolitics—and even of geography—has been closely linked to the history of Brazil and, particularly, to the needs of the State, remaining relatively isolated from the academic field. Geopolitics flourished in Brazil between the 1930s and 1950s, in a context of political and economic change, and resulted from the 'positivist' tradition that aimed at the economic and social modernization.[79] Geographical and geopolitical knowledge was considered a rational tool to promote the national development. The expansion to the interior, that is, to the Amazon Forest, was considered the path to achieve the territorial integration and to overcome the natural and social barriers of the tropical Brazil. But while geography remained in the hands of academics and universities, geopolitics were developed mainly by figures coming from other fields of expertise—and many times within the military.

After Brazil's involvement in the World War II, the international status of the country changed, and thus, it also did the geopolitical perspectives elaborated there. One of the main changes introduced was the adoption of a global scope in the analysis. While on the previous works, challenges for Brazil were posed by the natural characteristics and the potential territorial conflicts with its regional neighbours; during the Cold War, the main concern was defined by the global antagonism of the superpower competition and the disruptive potential expressed in internal destabilization and external aggression.

In that Cold War context, since the 1950s, an interest in Brazil's overseas projection of power was aroused. While interest in the Amazon remained constant, the innovation was the importance given to the western coast of Africa. In the logic of superpower competition, the South Atlantic was seen as a strategic line of defence against communist aggression. In that scheme, Antarctica was included with a marginal, but still relevant role in hemispheric defence.

Once again, after 1990, new changes on the international scene and Brazilian politics transformed the Brazilian geopolitical perspectives. The end of the Cold War, the redemocratization of Brazil, a deeper integration into the world economy and the beginning of South America regional

[79] Foresta (1992).

Table 1.1 Geopolitical thought in Brazil and the importance of Antarctica

	Economic context	*Historical context*	*Main concerns*	***Antarctica***
First wave of Geopolitical writings (Backheuser/Travassos) Influence from European authors	Transition from the primary exporting model to substitutive industrialization	Interwar instability and domestic unrest	Development of the interior to modernize and overcome the tropical nature of the country	Absent
Second wave of Geopolitical writings (Silva/Mattos/Castro) Influence from US authors and the Brazilian War College (ESG)	Developmentalist model of industrialization	Cold War logic of Superpower competition and internal conflicts, including military dictatorship	Continental defence and containment of social unrest. Economic development was important for the latter	Antarctica acquired some importance for national defence—due to its location—and the potential of alleged economic resources of strategic importance
Third wave of Geopolitical writings (Becker) Influence from globalization theory and environmental concerns	Neoliberalism and globalization	End of the Cold War and internal democratization	Regional development and environmental concerns	Absent

(continued)

Table 1.1 (continued)

	Economic context	*Historical context*	*Main concerns*	***Antarctica***
Contemporary writings	Limited industrialization in the context of the globalized world	Crisis of US hegemony and emergence of new powers in the international order	Revival of National security concerns: projection of power; South-South cooperation	Revival of the interest in Antarctica, albeit limited. Some revival of former 'strategic' concerns

Source The author

integration resulted in a reduced interest in traditional geopolitical issues. Those changes moved Brazil away from the previous 'National Security Doctrine' and induced it to embrace the process of globalization and the renewed emphasis given to the economic variables.

This process was reflected in the new Brazilian geopolitical writings that started to analyse the global geopolitical economic process—Becker's alleged renovation of the field that Kacowicz called 'geo-economics.' That change also signalled the introduction of universities and academics into the geopolitical production. The naturalist position of the former 'positivist' geopolitics gave place to critical perspectives, which were parallel to the critical reformulations of other fields, such as the constructivist turn in International Relations. Nonetheless, even those critical views were unable to overcome the state-centred perspective of Brazilian geopolitics.

As seen, the geopolitical writings in Brazil have covered a number of topics, adopting different scopes and perspectives. Only during the Cold War, with the adoption of a global scope in the Brazilian geopolitical analysis, Antarctica started to be included as a region of interest. However, in most cases, that interest was marginal, linked with its role as gateway of the interoceanic passage, the possibilities of its use on a global military confrontation and the hypothetical presence of strategic natural resources.

In contrast to Argentina and Chile, where Antarctica was seen as a natural extension of its territory and a space for conquering the frozen wilderness,[80] Brazil did not envision Antarctica as part of its territorial unit. For many Brazilian geopoliticians, the interest in Antarctica was mainly military and economic, but never above other more immediate interests either related to the country's interior or the South Atlantic. The instrumental role given to the Antarctic contrasted strongly with the essential function that the Amazon had. It was in the tropical Amazon Forest that Brazilian geopoliticians located its manifest destiny of exploration and conquest. As some geopolitical authors exerted a significative influence on the governmental circles, especially during the second half of the twentieth century, it is not surprising that Antarctica was considered a secondary issue and that Brazil's involvement in Antarctica adopted a rather subordinated character.

[80] On the constitution of a territorial Antarctic identity in Argentina and Chile, see: Cardone (2020).

References

Abdenur, A. E., & Neto, D. M. (2014). Rising powers and Antarctica: Brazil's changing interests. *Polar Journal, 4*(1), 12–27. https://doi.org/10.1080/2154896X.2014.913910

Anselmo, R. de C. M. de S., & Bray, S. C. (2002). Geografia geopolítica na formação nacional brasileira: Everaldo Adolpho Backheuser. In L. H. de O. Gerardi & I. A. Mendes (Eds.), *Do Natural, do Social e de suas Interações: visões geográficas* (pp. 109–119). Rio Claro: AGETEO/UNESP. Retrieved from http://www.rc.unesp.br/igce/geografia/pos/downloads/2002/geografia.pdf

Apontamentos Bio-Bibliográficos - Prof. Carlos Delgado de Carvalho. (1944). *Boletim Geográfico, II*(13).

Backheuser, E. (1952). *A Geopolitica Geral e do Brasil*. Biblioteca do Exército - Gráfica Laemmert Ltd.

Becker, B. K. (1986). Geography in Brazil in the 1980s: Background and recent advances. *Progress in Human Geography, 10*(2), 157–183. https://doi.org/10.1177/030913258601000201

Câmara, P., Barros-Platiau, A., Andrade, I., & Hillebrand, G. (2021). Brazil in Antarctica: 40 years of science. *Antarctic Science, 33*(1), 30–38. https://doi.org/10.1017/S0954102020000449

Cardone, I. J. (2014). A Política Antártica Brasileira no século XXI: evolução de fatores estruturais. In *Anais do Seminário Nacional Sociologia & Política*. UFPR.

Cardone, I. J. (2015). *As Posições Brasileiras No Sistema Do Tratado Antártico Com Ênfase Na Questão Ambiental*. Universidade Federal de Parana. http://hdl.handle.net/1884/38844

Cardone, I. J. (2020). Shaping an Antarctic identity in Argentina and Chile. *Defence Strategic Communications, 8*, 53–88. https://doi.org/10.30966/2018.RIGA.8.2

Carmona Gomes, R. (2012). *Geopolítica Clássica e Geopolítica Brasileira Contemporânea: Mahan, Mackinder e a "grande estratégia" do Brasil para o século XXI*. Universidade de São Paulo. Retrieved from http://www.teses.usp.br/teses/disponiveis/8/8136/tde-11062013-111229/pt-br.php

Carvalho, D. de, & Castro, T. de. (1956). A Questão Antártica. *Boletim Geográfico, XIV*(135), 502–506.

Carvalho, D. de, & Castro, T. de. (1960). *Atlas de Relações Internacionais*. IBGE. Retrieved from https://servicodados.ibge.gov.br/Download/Download.ashx?http=1&u=biblioteca.ibge.gov.br/visualizacao/livros/liv87309.pdf

Carvalho, J. M. de. (1997). Mandonismo, Coronelismo, Clientelismo: Uma Discussão Conceitual. *Dados, 40*(2), 229–250. https://doi.org/10.1590/S0011-52581997000200003

Castro, T. de. (1959). Antártica o Assunto do Momento. *Boletim Geográfico, XVII*(150), 238–245.
Castro, T. de. (1976). *Rumo à Antártica*. Freitas Bastos.
Castro, T. de. (2009). Carlos Delgado de Carvalho. In M. A. M. Santos (Ed.), *Geografia e Geopolítica: a contribuição de Delgado de Carvalho e Therezinha de Castro*. IBGE.
Child, J. (1986). *Antarctica and Argentine Geopolitical Thinking, 12*, 12–16. Retrieved from http://www.jstor.org/stable/25765660
Child, J. (1988). *Antarctica and South American geopolitics: Frozen Lebensraum*. Praeger.
Colacrai, M. (2003). Reflexiones en torno al régimen antártico y las relaciones argentino-chilenas. *Cuadernos de Politica Exterior Argentina. Rosario: Centro de Estudios en Relaciones Internacionales de Rosario (CERIR)* (73), 1–36.
Costa, P. C. (2009). Delgado de Carvalho: a trajetória de um educador. In M. A. M. Santos (Ed.), *Geografia e Geopolítica: A Contribuição de Delgado de Carvalho e Therezinha de Castro* (pp. 39–52). IBGE.
Costa, W. (1992). *Geografia Política e Geopolítica: Discursos Sobre o Território e o Poder*. HUCITEC/EDUSP.
Costa, W. (2012). Projeção do Brasil no Atlântico Sul. *Revista USP* (95), 9–22.
Dodds, K. (1997). *Geopolitics in Antarctica: Views from the Southern Oceanic Rim*. Wiley.
Ferraz, C. B. O. (2009). Delgado de Carvalho e a Geografia no Brasil: pioner-ismo e contribuições. In M. A. M. Santos (Ed.), *Geografia e Geopolítica: a contribuição de Delgado de Carvalho e Therezinha de Castro*. IBGE.
Foresta, R. A. (1992). Amazonia and the politics of geopolitics. *Geographical Review, 82*(2), 128–142.
Gandra, R. M. (2009). O Brasil e a Antártida: ciência e geopolítica. *Geografias, 05*(2), 65–74. Retrieved from http://www.jurua.com.br/shop_item.asp?id=12348
Gandra, R. M. (2013). *Geopolítica Antártica no Limiar do Século XXI: a definição de um projeto estratégico-científico para o Brasil na Antártida*. Universidade Federal de Rio Grande do Sul.
Henriques, E. de M. (1984). *Uma Visão da Antártica*. Biblioteca do Exército Editora.
Hepple, L. W. (1986). Geopolitics, generals and the state in Brazil. *Political Geography Quarterly*. https://doi.org/10.1016/0260-9827(86)90059-5
Kacowicz, A. M. (2000). Geopolitics and territorial issues: Relevance for South America. *Geopolitics*. https://doi.org/10.1080/14650040008407668
Karol, E. (2013). *Gografia Política e Geopolítica no Brasil (1982–2012)*. Universidade de São Paulo.

Kelly, P. L. (1984). Geopolitical themes in the writings of general Carlos de Meira Mattos of Brazil. *Journal of Latin American Studies*. https://doi.org/10.1017/S0022216X00007136

Landolfi, D. C., & Grynszpan, M. (1987). *Da revolução de 30 ao golpe de 37: A depuração das elites*. Retrieved from http://hdl.handle.net/10438/6583

Leal, V. N. (1948). *Coronelismo, Enxada e Voto*. Forense.

Mastrorilli, C. (1977). Brasil y la Antártida. La Tesis de Therezinha de Castro. *Estrategia* (43–44), 112–118.

Mattos, C. de M. (1977). *A Geopolítica e as Projeções do Poder*. Biblioteca do Exército Editora.

Mattos, L. F. de. (2014). A Inclusão da Antártida no Conceito de Entorno Estratégico Brasileiro. *Revista Da Escola de Guerra Naval, 20*(1), 165–192. Retrieved from http://www.egn.mb/arquivos/revistaEgn/nova-revista/revista-20n1/index.htm

Mattos, L. F. de. (2015). *O Brasil e a Adesão ao Tratado da Antártica: uma análise de política externa no governo Geisel*. Universidade Federal Fluminense.

Miyamoto, S. (1981). Os Estudos Geopolíticos No Brasil: Uma Contribuição Para Sua Avaliação. *Perspectivas* (70), 75–92.

Miyamoto, S. (1985). *A Geopolítica e o Brasil Potência*. UNESP.

Miyamoto, S. (1987). *Geopolítica e Política Externa Brasileira.pdf*. UNESP.

Penha, E. A. (2009). Geografia Política e Geopolítica: os estudos e proposições de Delgado de Carvalho e Therezinha de Castro. In M. A. M. Santos (Ed.), *Geografia e Geopolítica: A Contribuição de Delgado de Carvalho e Therezinha de Castro* (pp. 117–134). IBGE.

Sampaio, D. P., Cardone, I. J., & Abdenur, A. E. (2017). Brazil, the Antarctic Treaty System and Antarctica. In K. Dodds, A. D. Hemmings, & P. Roberts (Eds.), *Handbook on the politics of Antarctica*. Edward Elgar Publishing.

Sampaio Machado, M. (2009). A Contribuição de Delgado de Carvalho aos Estudos Geográficos Brasileiros a partir da obra "Le Brésil Méridional." In M. A. M. Santos (Ed.), *Geografia e Geopolítica: A Contribuição de Delgado de Carvalho e Therezinha de Castro* (pp. 53–62). IBGE.

Silva, G. do C. e. (1967). *Geopolítica do Brasil*. José Olympio Editôra.

Tambs, L. A. (1970). Latin American geopolitics: A basic bibliography. *Revista Geográfica, 73*, 71–105.

Travassos, M. (1947). *Projeção Continental do Brasil* (4th ed.). Companhia Editora Nacional.

Archival Sources

Antártida Argentina: ¿Qué pretenden los brasileños? (1976, February 25). [Clipping from an unidentified Argentine newspaper]. Antártida. – Dossiê (BR DFANBSB Z4.REX.IPS.27). Arquivo Nacional, Brazil.

Other Sources

Carlos de Meira Matos. Dicionário Histórico-Biográfico Brasileiro – DHBB. Source: http://www.fgv.br/cpdoc/acervo/dicionarios/verbete-biografico/carlos-de-meira-matos. Accessed on: 1 March 2018.
Decreto N° 1,200. (1936, November 17). Brazil.
Decreto N° 1,527. (1937, March 24). Brazil.
Decreto N° 24,609. (1934, July 6). Brazil.
Decreto-Lei N° 218. (1938, January 26). Brazil.
Golberi de Couto e Silva Dicionário Histórico-Biográfico Brasileiro – DHBB. Source: http://www.fgv.br/cpdoc/acervo/dicionarios/verbete-biografico/silva-golberi-do-couto-e. Accessed on: 1 May 2018.
Lei Federal N° 785. (1949, August 20). Brazil.

CHAPTER 2

Prelude to a Brazilian Antarctic Policy

2.1 Towards New Horizons

Brazil only joined Antarctic history in the second half of the twentieth century. Besides some collaboration with foreign expeditions, there was no Brazilian involvement on the southern latitudes before the 1950s. Not surprisingly, there is no evidence of any positive response from Brazil to the US Lt Matthew Fontaine Maury's letter of 10 April 1861, inviting Brazil to join an international effort to explore the southern latitudes.[1] The first contact Brazil had with polar endeavour was collaboration with the observation of the transit of Venus through the vessel *Parnahyba*,[2] the support provided to the Belgian explorer Adrien de Gerlache in 1898 and the assistance offered to the French explorer Jean-Baptiste Charcot in 1908. However, having consisted mostly of the donation of goods and other courtesies, these collaborations did not result in any direct

[1] Maury (1861).

The letter was also sent to representatives of Austria, Britain, France, Italy, Portugal, Spain, Russia and The Netherlands.

[2] In 1882, the vessel *Parnahyba*, under the command of Antônio Luís Von Hoonholtz, 'Barão de Teffé,' was sent by the Emperor Dom Pedro II to Punta Arenas (Chile), to collaborate with the studies of the transit of Venus that were taking place in simultaneous with the International Polar Year.

I. J. Cardone, *The Antarctic Politics of Brazil*,
https://doi.org/10.1007/978-3-030-80161-8_2

involvement with polar activities, and the impressions left by the expedition participants shows the little knowledge and involvement that the officials and general public of Brazil had in respect to Antarctica.[3]

Being focused on its territorial expansion to the west, where the tropical frontiers of the forest lay, Brazil's ambitions were directed towards the expansion of the monoculture economy and the harvesting of its natural resources. Economically, ideologically and, consequently, politically, the Antarctic region did not offer any particular interest to a country already characterized by a vast area of fertile land with infinite potential. By the turn of the twentieth century, no kind of trans-oceanic ambition was apparent between the Brazilian elites, and the idea of a territorially satisfied country crystallized as one of the principles in which it would design its foreign policy.

The establishment of the Republic in 1889 was followed by a brief period of instability in Brazil's foreign policy.[4] That changed when José Maria da Silva Paranhos Júnior, 'Barão do Rio Branco,' assumed the post of Minister of Foreign Affairs in 1902. Rio Branco reorganized the Ministry of Foreign Affairs, known as '*Itamaraty*,'[5] and restructured Brazil's foreign policy under well-defined principles. By standardizing the social origin of the Ministry personnel—bringing to the service men with family links to politics and the military hierarchies—Rio Branco established a *esprit de corps* that favoured the internal cohesion of the Brazilian diplomatic ranks.[6] His charisma, personality and influence paved the way to an increase of Itamaraty's influence and autonomy. That autonomy allowed the adoption of an encompassing view of the international relations strategy for the country. Up until then, each issue of Brazil's foreign

[3] Frederick A. Cook, surgeon and anthropologist of the 1898–1899 Gerlache's Antarctic expedition, expresses: 'This portrayed clearly the Brazilian notion of a polar expedition. The ideas proved to be so tropical that I must risk a breach of etiquette and quote enough to show Brazilian versions of polar work' (Cook, 1909, p. 7).

[4] Cervo and Bueno (2012, pp. 176–180).

[5] The name comes from the building the Ministry occupied in Rio de Janeiro, the *Palácio do Itamaraty*, when the capital of the Republic was located there. Presently, the building is still occupied by the Ministry's representation in Rio de Janeiro, while its headquarters are located in Neumayer's *Palácio dos Arcos* in Brasília, which is also known as *Palácio Itamaraty*.

[6] Pinheiro (2010, p. 11).

policy had been decided in isolation, without any principle to guide foreign action.[7] But Rio Branco changed that.

The minister's design for Brazil's international approach was oriented to solve the remaining border disputes with neighbouring countries through diplomatic channels, to switch the emphasis on the international alignment from Britain to the US and to search for a Latin-American alliance, with Brazil as the leader.[8] In Rio Branco's strategic view, all three options were intrinsically linked and offered Brazil a certain freedom of manoeuvre. The solution of the border disputes allowed Brazil to take advantage of a moment of relative stability in the international and regional systems and to consolidate Brazilian territory in front of all its Latin-American and European neighbours. The resort to diplomatic channels avoided unnecessary strain on Brazil's relation with the other Latin-American nations, in a moment when the Republic had allowed them to overcome the former enmity that resulted from the contrast between the republican and monarchic political ideologies. In doing so, Brazil took advantage of its favourable relations with the US, which was acquiring the role of Brazil's main commercial partner and political ally.

The 'unwritten alliance'[9] between Brazil and the US was underpinned by a pragmatic approach to international politics. By acknowledging the hegemonic role of the US, Rio Branco seized the opportunity of an alignment with that country, rather than resisting an alleged American imperialism. That orientation was reinforced by the increasing influence that the American market was gaining in the Brazilian economy. As the US gained importance for the Brazilian exports, the economic axis of the country switched from the northeast, based on the production of sugar cane, to the central-south, based on the production of coffee—the main product exported to the US.

That special relationship with the US was exploited by Rio Branco as a way to position Brazil as intermediary between that power and the Latin-American countries. Therefore, Brazil could benefit from US sympathy in any arbitration with its neighbours and project the image of leader within Latin-America. With that aim, Brazil proposed to multi-lateralize the

[7] Ricupero (2018, p. 406).

[8] Ibid.

[9] Bradford Burns (1966).

'Monroe Doctrine,' by promoting the idea of regional solidarity on hemispherical defence. However, that alignment with the US did not result in a complete submission, as Brazil maintained its relative independence when convenient.

Regarding Latin-America, the solution of disputes by diplomatic means and the approach after the constitution of the Republic were followed by Rio Branco's intent of becoming a regional leader. The US favour was exploited to exert influence over the other Latin-American countries. Pursuing a regional alliance, Rio Branco proposed an agreement between Argentina, Brazil and Chile that failed to succeed, mainly as a result of Argentina's reluctance to limit its political freedom. Nonetheless, Brazil's foreign policy during the years of Rio Branco signalled an attitude that would characterize Brazil's approach in the region: the intention to become the undisputable regional leader.

The abovementioned is relevant when interpreting several of the features that Brazil's policy for Antarctica adopted. The first characteristic was the technical character and autonomy that Itamaraty acquired, which helped to understand why the considerations of that Ministry predominated in the definition of Brazil's attitude towards Antarctica—even when other positions and interests were influential. Second, the idea of pacific and diplomatic solutions to disputes ended up constituting one of the principles incorporated as a cultural dictum within the diplomatic ranks and thus favoured international agreement over unilateral acts. Third, the pragmatic alignment with the US determined a non-confrontation policy, even maintaining a certain degree of autonomy, hindering any attitude that could embarrass the US or its allies. Last, the idea of becoming the Latin-American leader became a constant that extended to several fields of Brazil's foreign policy, including some proposals for a regional strategy for Antarctica.

Brazil's first overseas projection of power came when it entered the World War I and, subsequently, searched for a permanent seat on the Council of the League of Nations. After failing to obtain a permanent seat and its withdrawal from the League, Brazil became disenchanted with multilateralism. That had an impact on the way in which Brazil started to approach international organizations in the following years. However, what really shaped the way in which Brazil saw its role in the international system was defined more by the internal changes that were taking place in Brazil than any external factor.

In Brazil, important social, economic and political changes, accumulated during the first decades of the twentieth century and determined by the referred shift in the economic axis, strained the socio-political order based in the so-called Republic of the '*coroneis*' (chieftains or landlord). The federalist structure installed by the Republic gave too much power to local leaders, limiting the influence of the central power. The extremely limited form of political representation[10] and the usual resort to electoral fraud favoured the concentration of power in the figure of the local '*coronel*'—a sort of chieftain who based its authority on the inherited economic wealth and the traditional conception of social hierarchy.[11] But the advances in the process of urbanization and the massive immigration at the beginning of the century brought important changes to Brazilian society that started to exert pressures on the prevailing social order. Furthermore, the World War I favoured the development of incipient industries in the main cities due to the difficulties imposed by the war on the provision of industrial products from Europe, which increased the process of urbanization.

For the Brazilian *intelligenzia*, the ideas extracted from a so-called positivist ideology had inspired the formation of the Republic,[12] but was unable to overcome the limitations imposed by a society organized under a traditional order and a primary economy. Furthermore, on the part of the Brazilian elites, the so-called positivism was conservative in many aspects, and not necessarily directed to the promotion of scientific development. That expressed the characteristic tendency of the elites to adapt the philosophical traditions to their needs.[13] Nonetheless, the increasing urbanization of the population—even when very limited at that point—and the incipient industrialization set the scene for the incorporation of a new generation of intellectuals. Based on the ideal of modernization, those intellectuals promoted the technification and bureaucratization of

[10] The 1891 constitution granted universal and secret voting rights, but excluded the illiterate. Since this last group constituted the great majority of the Brazilian population, the resulting voting rights were, in effect, extremely restricted (de Oliveira & Pereira, 2016). This, added to fraud and forced voting (the so-called *voto do cabresto*), made political representation very limited.

[11] About the phenomenon of '*coronelismo*' see footnote 12 in Chapter 1.

[12] Foresta (1992).

[13] Schwartzman (1978, pp. 548–550).

all public spheres and provided the intellectual grounds over which a nationalist and centralist movement developed.

The social and economic changes accumulated during the interwar period led to the so-called Revolution of the 1930 that established a filo-fascist and authoritarian government led by Getúlio Vargas. Under Vargas, the Federal government initiated a policy of concentration and rationalization of organizational authority under the central administration. The reforms induced by the government tended to extract power from local authorities, making use of a more technical decision-making process as a way to prevent many of the States' policies from being politicized. These reforms gave rise to a set of technical and professional bodies that institutionalized and modernized many fields of state action.[14]

Moreover, it was also in the 1930s that a national science and technology system started to take shape. Previously, the educational system inherited from the Portuguese colonial structure limited scientific development within Brazil. Most higher education institutes were focused on law, medicine and engineering, and little emphasis was given to research.[15] The *Universidade de São Paulo* (University of Sao Paulo), in 1934, and the short-lived *Universidade do Distrito Federal* (Federal District University),[16] in 1935, were the first institutions to promote the development of a wider range of disciplines and to give focus to research. The influence of French academics at the Faculty of Philosophy, Sciences and Literature of the University of São Paulo was key to instilling that institution with a new academic ethos.[17]

These internal changes were combined with structural transformations in international circumstances: the world economic crisis that followed the 1929 Wall Street stock market crash and, afterwards, the rise of tensions in Europe that ended up with the outbreak of hostilities in September 1939. The world economic crisis cemented a change of the axis of Brazil's economy towards the US. On the other hand, it also evidenced the frailties imposed by the monoculture economic structure. The limitations

[14] For some details about the impact of that process on the field of geographical knowledge and the origin of geopolitical thought in Brazil, see Chapter 1 of this book.

[15] Up until the twentieth century, there were no higher education institutions denominated 'University' in Brazil.

[16] Schwartzman (1978, p. 557).

[17] Ibid., pp. 557–559.

imposed by the depressed international markets, both to the exportation of its primary products and to the import of industrial manufactured products, induced Brazil to adopt an economic model based on the development of an import substitutive industry. This phenomenon reinforced the increasing nationalism of the political spectrum, which found fertile ground in the masses who were displaced from the countryside to the cities.

The increasing urbanization and rapid industrialization of Brazil brought new types of technical challenges. Issues such as the rational analysis of the territory, the planning and implementation of national programmes of infrastructure, and the demographic analysis of the population started to be dealt with by the state apparatus. The rise of analytical geography, demographic studies and geopolitical thought accompanied that process, imbued by a positivist and rationalist philosophy. However, scientific research continued to be hampered by the weakness of the institutional setting of the Brazilian scientific system and the lack of a well-established scientific community.[18]

These complex structural and institutional changes did not alter any of the principles imposed by Rio Branco on Itamaraty. The position of neutrality adopted during the initial years of the World War II let Vargas maintain its favourable commercial relationship with the US while, at the same time, taking advantage of the German need of Brazilian products. However, as the conflict evolved and the US entered the war at the end of 1942, the maintenance of a neutral position became harder to sustain.

Despite some ideological affinities between the Vargas and the Nazi governments, the acknowledgement of the vantages than an approach to the US would bring to Brazil ended up inclining the balance in favour of the latter. As a result, the US started to invest in the development of heavy industry in Brazil in exchange for the support of Brazil in the war. After Germany sank two Brazilian commercial vessels in the Atlantic, Brazil finally declared the war on Germany in 1942. That decision helped not only to consolidate the US investment in heavy industry, but also to modernize and equip its military forces. In 1944, Brazil joined the allied forces that disembarked in Italy, securing in that way a place in the armistice negotiations and the formation of the United Nations (UN). Brazil's expectations to obtain a permanent seat in the UN Security

[18] Ibid.

Council did not come to fruition, but unlike the situation with the League of Nations, Brazil did not abandon the international organization, nor did it become disenchanted with multilateralism.

Vargas' authoritarian model had no place in a country that had sided with the allies against the Nazi-Fascism in Europe, and a process of liberalization concluded with Vargas' resignation and the call for general elections in October 1945. But Brazil's involvement in the war had changed the vision of the military, which started to push for the modernization of their equipment and training. Furthermore, the increasing tensions between the Western and the communist blocs made the existence of a modern and efficient military force important.

To that end, on 20 August 1949, the *Escola Superior de Guerra* (War College—ESG) was created. It was one of several institutions that emerged between the late 1940s and the 1950s which operated as a transmission belt between the Brazilian intellectuality and the implementation of their ideas through the state policies. In the case of the ESG, its creation responded to the request to provide an institutional setting to the ideology created within a noteworthy and influential part of the military leadership that interpreted the armed forces as the fundamental pillar against the dangers of the communist advance.[19] Inspired on the US National War College and the French *Institut des Hautes Etudes de la Defense Nationale* (Institute of Higher Studies of National Defence),[20] these intellectuals were heirs to the authoritarian and statist tradition that inspired the revolution of the 1930, but adapted to the context of the Cold War and the increased technical complexity of political decisions. One of the main concepts mobilized was the 'National Security Doctrine,' which included not only the geopolitical and military features of defence but also aspects linked to the economic and social systems. That is why their production covered a very wide variety of issues, being very influential between the military and the governmental circles in general.

Coeval with these developments, the scientific community established through the influence of foreign academics in Brazil started to push for further advances in scientific research. The creation, in 1948, of the *Sociedade Brasileira para o Progresso da Ciência* (Brazilian Society for the

[19] Dias de Oliveira (2010, pp. 23–25).

[20] Hepple (1986, p. 81).

Advancement of Science) marked a continuous effort to institutionalize the role of scientific institutions within Brazilian society and governmental circles. This was complemented, in 1951, by the establishment of the National Research Council (CNPq), which offered the first national state-based scheme for scientific research funding. Furthermore, following the interest that atomic physics generated regarding its strategic importance after World War II, in 1955, the Brazilian *Comissão de Energia Atômica* (Atomic Energy Commission) was created.

In addition to the Cold War context, these events were framed by a territorial competition in Antarctica, especially on the part of the Argentine, Chilean and British, who were involved in a complex balance of power scheme in the Antarctic Peninsula.[21] At the beginning of the twentieth century, with the development of the whaling industry, the interest in the economic resource established conflict in the Antarctic.[22] British interests in the region collided with the Argentine and Chilean views,[23] but conflict was avoided in order to avoid disturbing the commercial relations between those countries.[24] However, that changed with the advent of the World War II.

Several factors contributed to the Argentine and Chilean becoming more active in Antarctica. With the increasing involvement of European powers and the US in Antarctica, the call to an 'International Exhibition of Polar Exploration' to be held in Bergen, Norway, provoked a strong reaction in Argentina and Chile. Both countries were suspicious that the event was oriented to a division of Antarctica, and thus adopted measures to prepare to defend their respective positions in Antarctica, including the establishment of an Antarctic Commission in each country during 1939. While the Bergen exhibition was cancelled due to the war, the declining power of the British Empire was an opportunity that the two Latin-American countries could seize to advance on their position in Antarctica.

In 1940, Chile issued a decree defining the geographical limits of the Antarctic sector under 'Chilean sovereignty.' The Chilean decree angered

[21] For a throughout chronology and analysis of the conflict between these three parties, see: Fontana (2014) and Howkins (2008b).

[22] For a detailed account of that conflict, see: Cardone (2019, Chapters 2–3).

[23] The Norwegians were involved in the conflict at a certain point, especially because they were the main operators of whaling fleets (see, Cardone, op. cit.).

[24] Fontana (2014, p. 55).

the Argentines, but did not stop the two parties from holding negotiations to agree on a common policy[25] in front of extra-continental powers. Both countries had started to refer to the Antarctic sector south of the American continent as the 'American Antarctica'[26] or the 'American Quadrant,' defined as the sector comprising between about 25° and 90° west longitude. The concept had the political appeal of defining a continental sector against the pretension of any extra-continental power—particularly the United Kingdom—and was thus adopted by the Argentine and Chilean representatives in 1941 with the view of achieving an agreement regarding their respective Antarctic borders.[27]

Furthermore, to consolidate their respective positions in Antarctica, both countries developed plans to undertake a systematic exploration of the Antarctic region. The Argentines had been involved in Antarctica since the end of the nineteenth century, collaborating with several expeditions, including the rescue of the Swedish expedition of Otto Nordenskjöld, and operating, since 1904 and uninterruptedly, a station in the Laurie Island of the South Orkneys.[28] However, it would be in the 1940s that an extensive systematic programme of Antarctic exploration, including scientific work, the establishment of permanent and semi-permanent stations, and the performance of symbolic acts of sovereignty took place by the Argentines. For its part, Chile's active involvement in Antarctica was more sporadic but counted with the famous rescue of the Shackleton's *Endurance* Expedition in 1917. While Chile also planned to send an expedition in the early 1940s, it had to wait until 1947 for its first Antarctic expedition.[29] Nonetheless, they were able to send observers with the Argentine expeditions.

[25] The negotiations took place in 1941 and involved the President of the Argentine Antarctic Commission, Dr Isidoro Ruiz Moreno, and the representative of the Chilean Antarctic Commission, Dr Julio Escudero Guzmán. Previous negotiations took place in 1907 but arrive at no definitive understanding (Cardone, 2019, pp. 85–86, 215–216).

[26] Cardone (2020, sec. 3) and Howkins (2006).

[27] For the content of this preliminary talks, see: Copias de las Actas Firmadas en Santiago de Chile… (1941).

[28] In 1904, the Argentine government accepted to take over the meteorological and magnetic station established by the Scottish explorer William Spiers Bruce in 1903. At the time, no political significance was attributed to the fact by the British government, despite the worries it raised in Bruce himself (Swinney, 2001, 2007).

[29] The delayed entering of Chile into Antarctic exploration was a consequence of the limitation on resources, what was compensated by the symbolic impact of some of its

The Argentine and Chilean activities motivated a reaction from the United Kingdom. In 1943, using the activities of the Nazi raiders in the southern region as excuse, the British organized 'Operation Tabarin.' It was an operative aimed to establish permanent stations that could forestall any pretentions from Argentina and Chile and comply with the condition of permanent occupation. Once the war was over, the operation was reorganized under the Falkland Islands Dependencies Survey. The encounters between the British and Argentina and Chile varied from a formal interchange of protest to active hostility and required, from 1948, a declaration that put a hold on the dispatch of additional warships than those already in use for their ongoing operations south of 60° south latitude. However, the tripartite declaration did not stop the escalation of conflict that became critical in 1953.

With no other Latin-American country involved in Antarctic activities, Argentina and Chile promoted the idea of the 'American Antarctica' in opposition to extra-continental powers. To that aim, they introduced the issue in numerous Pan-American forums and pressed the negotiations of the 1947 Inter-American Treaty of Reciprocal Assistance—usually known as the Rio Pact, or TIAR from the Spanish acronym—to include the Antarctic sector lying between 24° and 90° west longitude within the definition of the continental zone of defence, even against the reservation made by the US in the contrary.[30] Despite the name given to the concept, the idea promoted by both governments was that such a sector belonged to Argentina and Chile, and that it was not extended to other countries in the region.[31]

In this context, it is not surprising that the possibility of expansion into an unknown continent—with its alleged economic and strategic potential—by two of Brazil's regional counterparts raised concerns in those who had been formed within the Brazilian tradition of geopolitical thought. That, and the increasing urge for overseas expansion that the involvement of Brazil in the World War II had incited, prompted the ideas of a young female geographer recently appointed as instructor of the ESG, Terezinha de Castro. In 1955, Professor Joaquim Ribeiro had conceded

actions, such as the visit of the Chilean President, Rafael Videla, in 1948, of a newly established Antarctic station, constituting the first visit of a Head of State to Antarctica.

[30] On early Argentina and Chile's relation with Antarctica, see: Cardone (2020).

[31] By that time, no other Latin-American country has expressed or demonstrated any interest in Antarctica.

in an interview that he defended the participation of Brazil in the decisions concerning Antarctica, based on historic, economic and strategic reasons.[32] However, it would be Terezinha de Castro's 1956 paper,[33] co-authored with her mentor, Delgado de Carvalho,[34] which would become a reference of the question in Brazil, being regarded by many as the first explicit study of the question.

The immediate reason for Castro and Delgado's paper was the announcement of the upcoming International Geophysical Year (IGY). The IGY had been organized since 1950 through the World Meteorological Organization and other scientific institutions, including the International Council of Scientific Unions.[35] It was originated as a Third Polar Year, but ended up widening its scope and incorporating the subtropical and tropical areas in order to study global geophysical phenomenon. The IGY was a massive international scientific event, which included some 30,000 people from 67 countries. While the event had a global scope and even covered outer space within its programme, great emphasis was given to the Antarctic, the most unexplored and unknown region of the world.

The IGY Antarctic programme eventually involved about 5,000 people from twelve countries, operating through fifty-six stations—including six sub-Antarctic stations. Despite it being a scientific event organized by scientists, the national governments were deeply involved in its inception and organization, due to the massive scale and resources needed. Moreover, notwithstanding the effort to keep political disputes away from the scientific collaborative effort, some conflicts prevailed—including the

[32] Ribeiro (1955).

Pinto Coelho (1983, p. 222) considers Ribeiro the first Brazilian in publicly speak about Brazil's involvement in Antarctica. As main arguments, Ribeiro defend the Brazilian right to an Antarctic sector based on 'historical reasons' derived from the Tordesillas Treaty, the 'economic reasons' derived from interest in the whaling activity and the presence of 'atomic minerals', and 'strategic reasons' consisting on the need of a defensive base in Antarctica for the control of the South Atlantic.

[33] Carvalho and Castro (1956).

[34] It is relevant to highlight that, at that time, Delgado de Carvalho was an instructor on the School of Diplomacy of Itamaraty, lecturing a course of World Diplomatic History (*Anuário do Instituto Rio-Branco 1956–1957*, n.d., p. 26).

[35] For a detailed account of the IGY, see: Barr and Lüdecke (2010) and Cardone (2019, sec. 5.1).

dispute between the mainland China and Formosa/Taiwan.[36] The discussion of the Antarctic programme was not absent from conflicts and, since the beginning, Argentina and Chile raised their concerns about the political aspects involved.[37] The solution was a so-called gentleman's agreement through which it was agreed that any action taken as part of the IGY would not be used for asserting or denying any claim to Antarctic territory.[38]

It is unclear if Castro and Delgado were aware of these discussions within the IGY Antarctic programme organization, but they seemed to confuse the nature and extent of the event. Misinterpreting the IGY—in which Brazil was taking part of the main program—as a 'Geophysical Conference' or 'Congress'[39] that would mimic the logic of the Berlin Colonial Conference of 1884–1885, the authors called for the urgent adoption of a national strategy to be included in the negotiations of the Antarctic continent.[40] Considering that the interests involved were not 'economic or related to tourism' but geostrategic, Castro and Carvalho contended that 'In the present moment, Brazil is in a situation that allow it to claim rights, which juridical basis are as strong as any other power.'[41] The crucial geopolitical reasons that urged the adoption of an active policy, as defined by the authors, were: (1) the strategic position of the Drake Passage, (2) the potential for the instalment of air control bases and (3) the prospects for climatic control from the South Pole. With that in mind, they proposed a strategy based on three actions: (1) take the initiative to a south hemispheric American conference by Argentina, Brazil and Chile based on the 'right of frontage', (2) participation of

[36] Sullivan (1962, Chapter 4).

[37] Cardone (2019, p. 5; 2020, pp. 253–255) and Cardone and Fontana (2019, sec. 4).

[38] This was a position defended by Argentina and Chile since the first IGY Conference on Antarctica in Paris (1955) but that was only accepted once that the Soviet announced they were to join the Antarctic programme (Cardone & Fontana, 2019, p. 307).

[39] Carvalho and Castro (1956).

[40] Castro and Carvalho were not alone in rising concerns about the political nature of the IGY. Argentina and Chile had joined the event with those same concerns. However, the Berlin Conference and the IGY were hardly comparable—the first being a political meeting and the second a wide international scientific event—and it is notable that Castro and Carvalho seemed to ignore Brazil's participation in the main programme.

[41] Carvalho and Castro (1956, p. 503, own translation).

Brazil on the Antarctic negotiations of the 'Geophysical Conference of 1957–1958' and (3) undertake, through its navy, the exploration and progressive occupation of the 'frontage territory,' following the example of Argentina.

Notwithstanding Castro and Delgado's call, Brazil's absence from the IGY Antarctic programme had several reasons. At that time, Brazil had a relatively backward scientific system. Only in 1960, with the creation of the *Coordenação de Aperfeiçõamento de Pessoal de Nível Superior* (Coordination for the Specialisation of Graduate Personnel)[42] and the *Fundação de Amparo à Pesquisa do Estado de São Paulo* (Sao Paulo Research Foundation) in 1960, a more comprehensive state-sponsored structure to scientific research was established, even when most of its limitations in terms of scientific independence and long-term sustainability remained.[43] Furthermore, Castro and Delgado's proposal was to direct Brazil's effort to the Weddell Sea area, one of the most inaccessible places that demanded better suited equipment and personnel, something that Brazil lacked. There were also political considerations regarding the possibility of Brazil's involvement in the Weddell Sea area, once Argentina was operating the permanent Belgrano station since 1955, the British the Halley Station and Shackleton Base since 1956, and the US the Ellsworth Station since 1957.[44]

Those political concerns and the absence of a more intense interest in the southern regions shaped Brazil's attitude to India's Antarctic proposal within the United Nations (UN). In February 1956, India proposed to include the 'Antarctic Question' within the provisional agenda of the UN General Assembly.[45] The proposal motivated the first registered involvement of Itamaraty on the Antarctic question, following a report by the, by that time, junior official, Lindolfo L. Collor.[46] He emphasized the

[42] The official translation of the acronym is: 'Coordination of Superior Level Staff Improvement.' Other common translations are: 'Coordination for the Improvement of Higher Level Personnel' and 'Coordination for the Improvement of Education Personnel.'

[43] Schwartzman (1978).

[44] The Ellsworth Station was eventually ceded to the Argentines once the IGY was over.

[45] About the India's proposal and its eventual withdrawal, see: Howkins (2008a) and Chaturvedi (2013).

[46] Ferreira (2009, pp. 116–117).

strategic importance of the region and recommended to favour the internationalization of the continent, due to the impossibility of permanent occupation. The Brazilian representation in the UN was thus instructed not to relinquish the criteria of effective occupation (*utis possidetis*) as a basis for territorial rights, despite that discovery and frontage principles could be accepted as 'inchoate' titles due to the special conditions of Antarctica. As an alternative, it was recommended to favour the continent's administration under the UN, following an agreement between the interested parties.[47] However, India's position failed to obtain the favour of the Third World countries involved in Antarctica—Argentina and Chile—and was emphatically opposed by the British. That determined the withdrawal of proposal in 1956 and again in 1958, without inciting any reaction from Brazil that preferred to avoid the issue.

In 1957, the question was studied at the ESG, suggesting that a more assertive attitude should be adopted. Thus, it was recommended that, as soon as possible, an official and public statement be made regarding Brazil's rejection of all claims in Antarctica, reserving its rights of freedom of access and to defend its 'interests.' That was supposed to favour the possibility to advance a claim when considered timely. It was also suggested that, as a temporary option, an international administration of the continent could be accepted if Brazil was a participant and it would not imply a renunciation to its 'future rights.'[48] Those views were approved by other official departments, but no such action was taken.

While Castro and Delgado's main interpretation on the IGY was misinformed, their concerns over the political aspects involved were not. As the IGY was coming to an end, the question of permanence in Antarctica arose. By the middle of 1958, the Soviets suggested extending the IGY for an additional year, resulting in the so-called year of International Geophysical Cooperation. With the US and the Soviet Union unwilling to withdraw their presence in Antarctica and the remaining conflict between Argentina, Chile and the UK, the situation required a definitive arrangement. By that time, the US has started consultation with the other 11 parties that had participated on the IGY Antarctic programme: Argentina,

Lindolfo Leopoldo Collor should not be confused his father, Lindolf Leopold Boeckler Collor, who was a prominent journalist and politician.

[47] Abreu (1974a, p. 2).

[48] Ibid., p. 3.

Australia, Belgium, Chile, France, Japan, New Zealand, Norway, Union of South Africa, Union of Soviet Socialist Republics and the United Kingdom, to participate in a conference on Antarctica. Thus, on 2 May 1958, US President Eisenhower invited those countries:

> ...to negotiate a treaty providing for freedom of scientific investigations and continuation of international scientific cooperation in Antarctica and ensuring that Antarctica be used for peaceful purposes only.[49]

The notice of the US invitation provoked the first official Brazilian position regarding Antarctica. Despite the high principles invoked in Eisenhower's invitation, the Brazilian government rejected the idea of accepting any agreement in which they had not participated that could affect a region in which it was considering holding strategical importance for national defence.The basic reasoning behind Brazil's official position was that the administration of Antarctica was linked to important strategic and economic interest, and that such interest was a legitimate basis for rights.[50]

In a confidential letter sent on 30 July 1958,[51] the Brazilian government rejected the criteria adopted by the US government of inviting the other participants of the IGY Antarctic programme as applicable to any other aim than the continuation of scientific cooperation.[52] In particular, the idea of an international agreement guaranteeing the 'peaceful use' was denounced under the premises that: '...due to geopolitical reasons, the proscription of Antarctica for military aims constitute a threat to national security to those nations that are not in the roster of nations invited to the pretended conference.'[53] Added to that consideration, the note highlights that the fact of having not sent an expedition to Antarctica during the IGY could not be considered a lack of interest, Brazil being the country with

[49] Coffman et al. (1991, p. 842).

[50] The *Realpolitik* of Brazil's geopolitical though seems to have been very present in such considerations. Regarding the official position, see: Visconti (1958).

[51] The letter remained confidential until Brazil's signature of the Antarctic Treaty in 1975. See: Ferreira (2009, p. 124).

[52] An abridged transcription of the protest note in Portuguese could be found in Rodrigues (1998, p. 325). A copy of the Original Note could be found in: Secretaria de Estado das Relações Exteriores to American Embassy, 1958.

[53] Ibid., own translation.

the most extensive coast in the South Atlantic and, thus, being exposed to threats that could come from those regions. Referring to the inclusion of an Antarctic sector within the region defined by the TIAR, it ended up asserting that, in front of a potential Soviet threat in Antarctica,[54] if the Conference was unable to reach an agreement that ensured the effective peaceful use of the continent, Brazil would reserve its freedom of access to Antarctica and to assert any claim that could be considered as necessary in future.

In November that year, the Brazilian Ambassador in Washington, Ernani do Amaral Peixoto, informed the Brazilian Secretary of State that the US remained adamant on limiting participation to the original 12 members of the Antarctic programme of the IGY, fearing that opening membership could favour the Soviet intention of including other countries of the communist sphere and could hamper the willingness of Argentina and Chile to sign the agreement.[55] Despite the Brazilian government's declared interests in the region, it seems that they were not considered priority enough to motivate further action.

By 1959, once the IGY ended and the political future of the continent was being discussed between the 12 nations involved, Therezinha de Castro published another paper,[56] which corrected many of her previous misconceptions regarding the IGY[57] and further developed her ideas for the inclusion of Brazil in Antarctic politics. Making use of the concept of 'American Antarctica,' the author defended the idea of the Antarctic Peninsula as a geological continuation of the South American continent. That use of the concept by Castro seems to have followed the original geographical and geological formulation by the Chilean Riso Patrón[58]

[54] The idea of a Soviet threat was also expressed in the media. See, for example: Guerra Fria entra em sua fase gelada (1957).

[55] Peixoto (1958).

[56] Castro (1959).

[57] However, the author felt in a geographical error, describing the continent as being separated '...by 4,000, 5,000 and 6,000 kilometres from the Tierra del Fuego, Tasmania and the Cape of Good Hope, respectively' (Castro, 1959, p. 245). The actual distances are about 1,000, 2,800 and 4,000. It seems that Castro referred to the distance to the geographical south pole instead of the distance separating the two land masses. It could be the case that, highlighting a longer distance between the countries, Castro was relativizing the closer proximity of Argentina and Chile to Antarctica, despite at this point that possibility is purely speculative.

[58] Riso Patrón (1908).

and not the political interpretation defended since 1941 by Argentina and Chile. Considering the 'Congress' (she was referring to the IGY) had 'nothing political,' nonetheless, Castro insisted on the strategical value of the Drake Passage, the need of bases for the air control and the utility of Antarctica for climate-related issues. She also considered that the exploitation of economic resources was still a possibility there and reminded that also in Africa '...it first prevailed the scientific spirit.'[59]

Continuing with the idea advanced on the 1956 paper, she defended the division of the continent based on the projection of the extreme limits of the coasts 'confronted' with the Antarctic, what collided with the already advanced claims of Argentina, Chile and the UK, as shown in Fig. 2.1. Notwithstanding the evident conflict with the Argentine and Chilean claims, as a final call, she urged: 'Let's claim our share of Antarctica! If they appeal to history, lets appeal too. Let's join to the South American countries Argentina, Chile, Ecuador, Peru and Uruguay around the right of frontage or Polar Sectors...'.[60]

However, there was little that Castro's call could achieve at that point, and onl December 1959, the Antarctic Treaty (AT) was signed without Brazil's participation.[61] Negotiations for the Treaty were lengthy.[62] A first round of preparatory meetings extended for more than a year, arriving at a draft of the agreement that was to be discussed on the Conference of Antarctica. The Conference started on 15 October 1959 and was characterized by an intense process of bargaining in which the definitive Treaty took shape. The AT granted the demilitarization of the continent, reserving it for peaceful purposes only and promoting international cooperation in scientific research on behalf of all humankind. Its Article IV

[59] Castro (1959, p. 243).

[60] Ibid., p. 245.

[61] There were previous attempts to reach an Agreement regarding Antarctica. Since the 1940s, Argentina and Chile called the need for an international conference, and in 1948, proposals were advanced by the United States and Chile but they were unsuccessful (Cardone & Fontana, 2019). However, the Chilean proposal included the stand-still idea that would result in the 'gentleman's agreement' of the IGY and the Art. IV of the Antarctic Treaty. The British unilaterally submitted the dispute about the Antarctic Peninsula area to the International Court of Justice in 1955 but, as Argentina and Chile rejected the jurisdiction of the court, the case was dismissed in 1956.

[62] On the origins and negotiation of the Antarctic Treaty, see: Beck (1985), Cardone (2019, sec. 5.3), Cardone and Fontana (2019), Hall (1994) and Peterson (1988, Chapter 2).

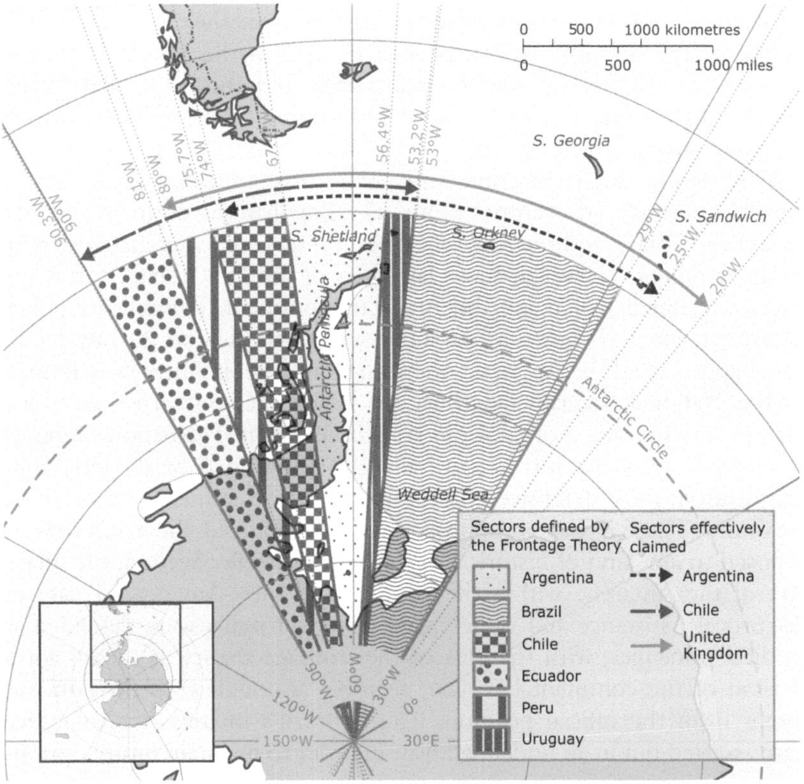

Fig. 2.1 Map of the division of the Antarctic Peninsula proposed by Castro's frontage theory and claims advanced by Argentina, Chile and the UK (*Source* The Author, based on '*Atlas de Relações Internacionais*' [Carvalho & Castro, 1960, p. 159])

froze all territorial claims, providing a formula that allowed the agreement without compromising anyone's position. It also prohibited nuclear testing and established a system of mutual inspections and consultations that ensured the objectives of the Treaty could be fulfilled. In order to avoid disagreement, some clauses of the Treaty remained ambiguous, and some aspects remained undefined—as the issue of natural resources. In terms of the possibility of accession, the Treaty remained open to any member of the UN or its specialized agencies. Nonetheless, active participation in the decision-making process defined in the Article IX of the Treaty was restricted to the original parties and those that have been accepted by consensus after having demonstrated '…substantial scientific research activity…' in Antarctica.

Considering the 1958 confidential note was sufficient guarantee for Brazil's interests and eventual rights, the Brazilian government did not protest the Treaty, and the Antarctic question was once more forgotten by the government. However, those last years of the 1950s shaped in many ways how Brazil would see the Antarctic continent, defining its policy towards it. First, as shown in the letter sent to the US on 30 July 1958, Brazil approached the Antarctic question from a geopolitical view framed on the National Security Doctrine, for which the southern region was strategic in the context of the Cold War competition. The national interest was defined by some sort of natural right determined by its geographical position: most extensive coast, meteorological influence and relative proximity. Thus, it was an approach that emphasized the *realpolitik* as opposed to any kind of historical or juridical principle. Second, the inclusion of the Antarctic within the zone of the Inter-American Treaty of Reciprocal Assistance and some vague historical rights were defended as juridical principles, with the idea of the frontage theory as a basis for a division of the continent between 'sectors,' although such division was absent from the official position. Third, Brazil's interests in Antarctica were asserted but in an undefined way, in order to maintain future options open.

The Brazilian official attitude regarding Antarctica at the time of the signature of the AT in 1959, reserving its rights and protesting against the Conference but not implementing any positive action, could be interpreted as a product of its relative lack of specific interest in the region. While some vocation of overseas projection of power, national security considerations and the regional power-balance competition prevented Brazil from accepting an international regime without its participation, the willing to not antagonize the US and its South American neighbours, the relative lack of interest and, particularly, the shortcomings in terms of operational polar capabilities had prevented Brazil to adopt a more assertive attitude. In contrast to its regional neighbours, Brazil didn't have diplomats with experience in Antarctic politics, it hadn't had any involvement in the whaling or sealing industry that could lobby the government, it lacked any polar tradition, training or equipment and it had more immediate and urgent political concerns to direct its energies and efforts.

2.2 The First Brazilian Initiatives

Brazil did not take part of the Antarctic programme of the IGY, but it was during that event that the first Brazilian set foot on the white continent. In early 1958, Anésia Pinheiro Machado, a famous Brazilian female pilot, took part in a tourist trip to Antarctica organized by the Argentine Navy. She was the first registered Brazilian national to set foot in Antarctica, although, as it has been usual in Antarctic historiography, her feat was disregarded. That recognition was attributed to a male Brazilian, who arrived in Antarctica some weeks after Machado and another Brazilian woman.[63]

While the female presence of Brazil on the continent went mostly unnoticed, the participation of the physician Professor Durval Rosa Borges, sent as a reporter from the newspaper *Correio da Manhã* and the *Visão* magazine on the US expedition Deep Freeze III in the same year, resulted in a number of articles and the subsequent publication of a book.[64] This led to the acknowledgement of Borges as the first Brazilian to travel and set foot in Antarctica.[65] The traveller was portrayed as the '...first Brazilian scientist to travel to and study the Polar regions,'[66] although the US invitation was granted to Borges as reporter and he did not participated in any research of the IGY. Borges' Antarctic experience resulted in a trip around the world, so his impressions about Antarctica were mixed with others coming from other latitudes, resulting in an account in which Antarctica was merely one—even when the most relevant—of several stops. His favourable view of the way in which the

[63] Within the limits of the present research, Machado was the first Brazilian national to travel to the continent. Apart from some reports in the press, there is little—if any—information regarding her role as the first Brazilian in Antarctica. See, for example: Foi para a Antártida (1958). Another female Brazilian is registered in the second trip of the Argentine Navy organized tourist trip to Antarctica, Erika Banr. She would be the second Brazilian national in visit Antarctica, before Borges did.

[64] Borges (1959).

[65] This acknowledgement is prevalent in the literature.

[66] Embarcou para o Pólo Sul (1958).

human withstood the hardness of the Antarctic environment, the international collaborative spirit that existed in the scientific work developed in Antarctica and the interchange between scientists in international forums did little to raise the public and official interest in Antarctica, notwithstanding his astonishment that Brazil was not taking part in any activity on such an important continent.

Between 1958 and 1961, some Brazilian nationals embarked as press reporters or cinematographers in expeditions of other countries without getting much notice.[67] However, in 1961, the issue gained public attention thanks to the participation of the meteorologist Rubens Junqueira Villela on the US Navy Operation Deep Freeze. Villela went to Antarctica in January, remaining there until April. This long stay let the meteorologist conduct several studies and to have relevant experience of the Antarctic work conditions and its scientific possibilities. The visit was followed by a second invitation later that same year. During that expedition, in November 1961, Villela reached the South Pole by plane, becoming the first Brazilian to reach the southernmost latitude.

Villela's participation in the expeditions was formalized by his designation as observer by the CNPq. Filling that role, he published an account of his observations on the Bellingshausen Sea. In that report, he concluded that such work had a direct practical application to foster the possibilities of weather forecasting in South America.[68] In an article, which appeared in the *O Estado de São Paulo* newspaper,[69] Villela also acknowledged the impressive logistical capabilities of the US operations, highlighting the upcoming inauguration of a nuclear reactor in the McMurdo Station[70] and the installation of the 'Ski-Hi' station. In his view, the station was strategic, as its location was at the centre of a potential air-route between South America and Australia.

In the years that followed, the only Brazilian scientists to go to Antarctica were the oceanographers Ibiracy Minussi and Adilson Murillo from the Oceanographic Institute of the University of São Paulo. These

[67] Brasil Precissa Descobrir a Antártida (1971).

[68] Villela (1962).

[69] Meteorologista brasileiro que foi ao polo sul (1961).

[70] Despite the Antarctic Treaty had banned nuclear testing in Antarctica, it did not prohibit the use of nuclear energy. The reactor operated for ten years, but, due to leakages, it was decommissioned. Afterwards, it was removed, jointly with the contaminated land (Musto, 2017, pp. 26–27).

two scientists took part of the 1963 Deep Freeze expedition, once more, by invitation of the US government.[71] That same year, Lieutenant Commander Sálvio Augusto de Oliveira Martins participated as observer in the Chilean Navy Antarctic expedition, becoming the first Brazilian military to visit the white continent.[72] In subsequent expeditions, Lieutenant Commanders Mario Edelman and José Henrique França Gomes were likewise offered a place as observers on the Chilean Antarctic expeditions, a practice which ended in 1967.[73]

Following the increasing public interest in the issue, on 18 March 1964, the Federal Representative, Cunha Bueno, presented a request of information to Itamaraty regarding the position of Brazil in the Antarctic. Such query was divided into three topics: (1) if the Ministry had any study that definitively suggested that Brazil should stay from of the Antarctic issue; (2) if the non-participation of Brazil on the signature of the AT meant a disinterest by Brazil on the continent and its international administration; (3) if the National Security Council had any evaluation regarding studies produced by the ESG that demonstrated the vital interests of Brazil in the continent.[74]

On 29 May of that year, the Ministry of Foreign Affairs responded to the three questions by considering that: (1) the studies developed by the Ministry suggested the accession of Brazil to the AT due to the advantages of constituting a concrete manifestation of interest by Brazil without compromising an eventual claim of interest and ensuring a place on the deliberations regarding the future of the Continent—even acknowledging that only the twelve original signatories and those demonstrating an active Antarctic programme could actually had a say on the regime; (2) that in 1958 the Ministry asked to be included in the negotiations for the Treaty, which was denied by the US government and resulted in a note of protest reaffirming Brazil's interest and the reservation of freedom of access; and (3) that such enquires should be submitted to the National Security Council instead.

[71] Brasil Precisa Descobrir a Antártida, op. cit.

[72] Mattos (2015, p. 123).

[73] Ibid.

[74] Request of Information No. 1520/1964, cited in: Brasil Precisa Descobrir a Antártida, op. cit.

However, Bueno's intervention in the Congress was followed some weeks later by the military coup of 1964 in which the elected President Jõao Goulart was overthrown by the Armed Forces in collusion with the US government and opposition forces. The coup was the end of a series of political incidents that started with the election of the leftist candidate Jânio Quadros and resulted in the direct control of the government by the military, who installed a tutelary authoritarian regime, characterized by a tight system of control of civil society and the political system. Apart from the installation of a highly authoritarian and violent system of political repression, the regime pursued a conservative cultural and religious agenda, and the rapid development of the industrial system. In the context of political and ideological repression and the fierce economic reform imposed on the Brazilian economy, it is hard to imagine that Antarctica could have occupied any sort of relevant place within the governmental priorities.

Thus, the question went into abeyance until 1970, when a report by the Joint Chiefs of Staff (EMFA) seemed to depart from the idea of a territorial claim established by the frontage 'theory' and recommended that:

> For a country such as Brazil, with such a great geographic expanse, it seems that it would be of utmost convenience the dispatch of Brazilian technicians and scientists to the Antarctic region—geologists, meteorologists, geographers, physicists, physicians, biologists, and specialists from our armed forces—with the intent of joining the expeditions and stay at the stations already established, all within the programme of international scientific cooperation established by the Antarctic Treaty.
>
> Brazil could easily join by a simple diplomatic procedure...[75]

Notwithstanding the recommendation of the EMFA, just a few months later, the Federal Representative Euripides Cardoso de Menezes adopted the opposite view in the Congress. Based on the arguments elaborated some fifteen years before by Castro, in the House of Representatives, Menezes appealed to the government to claim a sector of Antarctica based

[75] Estado-Maior das Forças Armadas (1970, p. 15, own translation).

on the 'frontage principle,' including the sector defined by the geographer on her earlier work.[76] The following year, in an article published in the magazine *Manchete*,[77] Menezes expressed consternation at the indifference with which the white continent was treated within Brazil and compared that attitude with the lack of vision of those who criticized the US Secretary of State when proposing the acquisition of Alaska. In furthering his position, Menezes defended that those who control the Poles would control the world. In his view, Antarctica held an important quantity of uranium and enough supplies to feed the whole world 'for centuries.' As a first step for Brazil, he proposed to establish a meteorological station, thinking it would require few resources but would hold important political significance.

The revival of Castro's ideas through Menezes' lobby in Congress helped to mobilize some initiatives within the government, including a report by the EMFA, required by the Head of the Military Cabinet, João Batista Figueiredo, in the first quarter of 1971. The report restated the strategic importance of Antarctica due to the possibility of the installation of enemy bases, the control of the Drake Passage, the link between Antarctica and Brazil's climate, and the economic prospects derived from hydrocarbons, coal and uranium.[78] As a final recommendation to the President, the report suggested avoiding any recognition of Antarctic territorial claims by foreign countries and to promote, through the CNPq, Antarctic studies, taking advantage of the opportunities to interchange scientific data and participate in foreign scientific expeditions to the region.

However, not everyone in Itamaraty shared that view. In June 1971, in a report elaborated by the special advisor João Frank da Costa, he

[76] Mattos contends that Castro had a strong influence over Menezes' views on the Antarctic question (Mattos, 2015, pp. 164–167) and even trace back Castro's work 'Rumo à Antártica' (1976) to a previous version elaborated in collaboration with Menezes, but never published (ibid., pp. 166–167). Menezes Introduction in his work of 1982 also indicates a close relation between his ideas and the influence of Castro's elaborations (Menezes, 1982, Chapter 1).

[77] Menezes (1971).

[78] The Special Report (N°14 FA-2-145 of 16 April 1971) is described in Mattos (2015, pp. 171–172). Also, some of its conclusions are described in: Abreu, op. cit., p.5 and in Secretaria Geral do Conselho de Segurança Nacional (1974).

concluded that the AT was a consolidated fact, and there was no particular benefit for Brazil in acting against such an international framework.[79] Against the position defended by Castro and Menezes and contradicting the recommendation of the EMFA, Costa favoured the accession to the AT as the first step to be adopted. Such action would allow Brazil to operate in accordance with the international agreement and conduct scientific research, which would be necessary to achieve consultative status, as provided by Article IX of the Treaty. Yet, frictions with Argentina regarding the use of the rivers of the Parana basin made the Brazilian government postpone any effective action.[80]

Regardless of Itamaraty's position against operating outside the framework of the Antarctic regime, the political impulse given by Menezes was backed by the initiative of a group of scientists gathered at the Engineering Club of Rio de Janeiro, who started to organize a private scientific expedition to Antarctica. As early as December 1971, the club was considering the possibility of organizing an Antarctic campaign in informal and closed meetings that, following Pinto Coelho, were being 'observed' by the intelligence agencies.[81] The Engineering Club appointed a Working Group, eventually named the Organizing Committee for the Antarctic Programme (*Comissão Organizadora do Programa Antártida*—COPRAN), that met during 1972 and developed a general scheme, including the creation of the *Instituto Brasileiro de Estudos Antarticos* (Brazilian Institute for Antarctic Studies—IBEA), which was founded on 7 September 1972.[82] In the words of its President, João Aristides Wiltgen, the IBEA was composed by a '...prominent group of engineers, members of the War College graduate association, men of science and Brazilian intellectuals...' and constituted a '...civil association, apolitical, with exclusively scientific-cultural ends, of a national scope and with

[79] Ferreira (2009, pp. 124–125).

[80] Abreu, op. cit.m pp. 5–6 and Secretaria Geral do Conselho de Segurança Nacional, op. cit., p. 2.

[81] Pinto Coelho (1983, p. 224).

[82] Wiltgen (1977, p. 324). The date is significative as it coincides with the 150th anniversary of the Brazilian independence, showing the nationalist character that imbued the founders of the institute.

Do not confuse with the International Biomedical Expedition to the Antarctic, a SCAR Working Group on Human Biology and Medicine proposal aimed to study the effects of the Antarctic environment to the human body.

non-profit objectives.'[83] The directory of the Institute was constituted by prominent figures from different social substrates, including names already associated with Antarctica, such as Rubens Villela and Therezinha de Castro.

The IBEA was promoted between the national scientific associations and gathered support from some prominent individuals within the scientific and political circles. By October 1972, the IBEA had already obtained several commitments from companies and associations, including a donation from the director of a construction company of a prefabricated house to be used as an Antarctic base, the contribution of 90,000 cruzeiros (about USD15,000 of that time) by the *Federação Brasileira das Associações de Bancos* (Brazilian Federation of Bank Associations) and the provision of supplies and their transportation to the required location by the *Federação de Indústrias do Estado de São Paulo* (São Paulo Industrial Federation—FIESP), between other contributions in kind and funds. Such support was obtained thanks to the active campaign exerted by its founder members and from wide press coverage.

In order to obtain the public's favour, the IBEA usually portrayed Antarctica in the media as holding untold riches and exerting fundamental influence over meteorological aspects, affecting Brazil's economic activities. That justified both the need for scientific data and the interest in participating in the administration of the continent. Despite the alleged apolitical character of the organization, many within the IBEA defended the Brazilian involvement as a strategy directed to a future claim to Antarctica.

However, not everyone in the IBEA favoured the territorialist position expressed in the frontage theory and defended by Menezes. Dr Aristides Pinto Coelho, director of Science and Technology of the IBEA, warned about the difficulties that the frontage principle would confront against the position already attained by the countries of the Northern Hemisphere, who would be left with nothing in the division of Antarctica following the frontage principle. Instead, he expressed his faith in the '...contribution of science and technology.'[84]

Pinto Coelho's publicized position seems to have led to some discontent within the Engineering Club, especially between those identified with

[83] Wiltgen, op. cit.

[84] Defrontação não é a melhor rota (1973).

Castro and Menezes' position. Even with those differences, by 1973, the IBEA was able to obtain substantial support from companies and the Brazilian Navy. However, the internal divisions, what was perceived as an obstruction from the government, and the failure in acquiring a polar vessel seemed to threaten the viability of the expedition. That being so, by February 1973, Villela, who was a prominent member of the IBEA and by that time was considered an authority in Antarctic issues, expressed his scepticism about the expedition.[85]

To promote and publicize the work of the IBEA and its programme, in September 1973, the Organizing Committee prepared a conference covering Antarctica and the Brazilian territorial waters. The conference included speeches on the planned IBEA 1st Brazilian Scientific Expedition to Antarctica, the scientific aspects of the continent and the relations between Antarctica and Brazil.[86] In São Paulo, the support of Villela allowed the IBEA to formulate a relevant scientific programme in the waiting to be implemented. The Industrial Federation and other agencies assured their support of the upcoming expedition and provided a forum within which to promote the idea through conferences and seminars. Those events were used by the IBEA to stress the economic possibilities that a better knowledge of the continent could open up. Between the possible benefits to be accrued, the IBEA members listed the potential for fishing, farming, textile industries and retail.

The IBEA-proposed scientific programme included contributions from researchers from São Paulo, Rio Grande do Sul and Rio de Janeiro. The programme included 9 research lines: (1) oceanographic studies on primary productivity, salinity, temperature variation, plankton and benthos, etc.; (2) isotopic relation on water from rain and ice samples from Antarctica; (3) meteorological observations along the route to build synoptic meteorological charts; (4) isotopic relations on soil from Antarctic samples; (5) zoological studies to establish the ecological chain of Antarctica and the catch of specimens to be exhibited in museums and zoos; (6) study of the iodine cycle in Antarctica; (7) study of the *Chaenocephalus aceratus*[87]; (8) radioactive contamination along the route

[85] Expedição brasileira à Antártida (1973).

[86] A detailed programme of the conference can be found in: Pinto Coelho (1983, pp. 228–229).

[87] Known as the 'icefish,' a rare species that lack haemoglobin.

from rain, sea and ice water samples; and (9) chemical and bacteriological study of Antarctic ice for the determination of its potential use for human consumption. As designed, almost all lines of research were linked with some potential benefit to be accrued for Brazil. However, and despite the usual projected image of Antarctica as a reservoir of mineral richness, it must be noted that the programme included little in the way of assisting to determine the potential for mineral exploitation of Antarctica.

Negotiations between the IBEA and the US Representation to facilitate the use of an American vessel *Eltanin* failed to advance but initiated a relationship that would enable the IBEA to establish its first international presence. Two representatives from the IBEA were invited by the cultural aide of the US Embassy, Donald Reynolds, to participate in a seminar of the US Antarctic programme, organized by the Office of Polar Programs of the National Science Foundation. In contrast to the required low-profile of what was a 'taboo' issue in Brazil, the Brazilians expressed satisfaction on being able to talk openly about their plans and projects.[88] But the main outcome obtained during that visit was the invitation made from the US Antarctic programme to participate on its expedition for the 1973–1974 season.

Despite support from several quarters, the difficulties confronted by the IBEA in putting together the resources necessary for the expedition—particularly relevant for the acquisition of a vessel suitable for ice navigation—seem to have led some politicians to look for parliamentary support. In March 1973, the Senator Vasconcelos Torres—who was himself a member of the IBEA—proposed a bill to authorize the Executive to provide logistical support to the Engineering Club of Rio de Janeiro (the mother organization of the IBEA).[89] It took more than a year for the proposal to get the approval of the Justice and Constitutional Commission and the National Security Commission of the Senate, only to be held up by the Foreign Affairs Commission and archived in March 1975. The lack of support from the government came from the

[88] Pinto Coelho (1983, p. 236).

[89] The original proposal was the Senate Law Bill Proposal n° 7 of 1973 presented on 23 March 1973.

reluctance to antagonize the Argentines, at a time when the two countries were clashing due to a joint project between Brazil and Paraguay to establish the Itaipu dam in the Parana River, part of Del Plata basin.[90]

The understanding between Brazil and Paraguay regarding the idea of a joint project for using the Parana River for the generation of energy dated back to 1964, but it was in 1973 that a specific agreement to advance the project for an hydroelectric dam was signed.[91] The agreement between those two countries alarmed Argentina, not only because the latter had its own project for installing a dam on the river—which would be affected by the joint venture of Brazil and Paraguay—but also for geopolitical considerations. Argentina was worried that an alliance between Brazil and Paraguay could bring the geopolitical balance too much in favour of Brazil, who, comparatively, had advanced its industrialization and economic growth at a much greater pace than Argentina over the preceding decades.

By announcing the construction of its own energy project, the Corpus dam, Argentina blocked the advance of the Itaipu project, which led to intensive diplomatic action, and was partially successful in forestalling the implementation of the Brazilian-Paraguayan plan. With Brazil and Argentina on an intense bilateral conflict due to the Itaipu project, there was little interest from the Brazilian government in adding extra strain by supporting an expedition to Antarctica.[92] One of the main hesitations was that, despite its alleged purely scientific aims, the enterprise was linked with 'territorialist' ambitions derived from the frontage theory and was already being reported in Argentina as a further example of Brazilian 'imperialism.' Those concerns were added to the unstable political situation in Argentina.

The former Argentine President Juan Domingo Peron was returning to the country after almost twenty years in exile.[93] He was prevented from participating as a candidate in the elections, but his Party won the

[90] Secretaria Geral do Conselho de Segurança Nacional, op. cit., pp. 2–4.

[91] See: Fajardo (2004), Nickson (1982) and Rosenbaum (1973), among others.

[92] Following the accounts of Azambuja, the IBEA director, Wiltgen, was even 'invited' to Brasília and was asked to give up his plans for the Antarctic expedition (Azambuja, 2005, p. 30).

[93] He fled the country after the so-called *Revolución Libertadora* (Freedom Revolution) of 1955, which was actually a military coup followed by the installation of an authoritarian military dictatorship.

Presidency using a puppet candidate, Héctor Cámpora,[94] and thanks to a fragile alliance between its traditional members and the left-winged radical elements. Those two fractions of Peron's base had collided since the very beginning of Cámpora Presidency, making the contradictory views that each one held on what the movement was apparent. The conflict ended with the renouncement of Cámpora and his Vice-President, Vicente Solano Lima, and the designation as provisional president of the Chamber of Representatives, Raúl Alberto Lastiri, who was close to the President and belonged to the right-wing of the movement.[95] The eventual election of Perón as President did not appease the conflict, and the left-wing of the Peronist movement started a direct confrontation with the government. Thus, the volatile situation in Argentina presented an indecipherable panorama that recommended caution not to provoke any unnecessary strain.

At the end of December, the IBEA Director of Science and Technology, Aristides Pinto Coelho, participated in the US expedition, fulfilling the promise made a few months before. The scientist was sent in relative secrecy to Buenos Aires to embark on the ship that would call at Ushuaia before heading to Antarctica.[96] The expedition on board the *Hero* visited Deception, Livingston, King George (*25 de mayo/Vaterloo*) and Elephant Islands, during which the IBEA scientist was able to collect algae, fish, krill, rocks, fossils and samples from sea water and ice. Although some criticized the effort, the works undertaken were celebrated in some circles, and Pinto Coelho was invited to give speeches at several universities and other institutions. However, there were no further involvement in any other expedition and, despite being characterized by Pinto Coelho as the 'IBEA expedition to Antarctica,'[97] such

[94] To what extent Campora acted as President being a mere puppet of Peron is something open to debate, but the whole campaign was based on that idea, made evident by the slogan 'Campora a la Presidencia, Perón al Poder' (Campora to Office and Peron to Power).

[95] The designation of Raúl Alberto Lastiri was pushed through the constitutional procedure by embarking the President of the Senate—the second on the succession line—on an imaginary mission to Argelia.

Lastiri travelled to the Argentine Antarctic Base Marambio during his brief Presidency, on 10 August 1973.

[96] All the details and accounts of the trip presented here come from Pinto Coelho's work (1983, Chapter 6).

[97] Ibid., p. 237.

limited participation did not fill the prerequisite of 'conducting substantial scientific research activity' as required by Article IX of the 1959 AT.[98]

However, the creation of the IBEA and the advances made in establishing the Brazilian presence in Antarctica led the government to give a more serious thought to the question. Before the possibility of embarrassing the government by acting outside the ATS, the IBEA forced Itamaraty and the Executive to consider acceding to the Treaty as a way to operate within the international institutional agreement.

Additionally, the oil crisis of 1973 and the Antarctic oceanic drilling research undertaken by the US *Glomar Challenger* expedition had raised the profile of the potential Antarctic economic resources. The embargo on oil imposed by the Arab countries resulted in an increased interest in the potential of Antarctic hydrocarbons from economic and strategic points of view. The increase in the price of oil signalled the possibility of making Antarctic exploitation economically viable at the same time that the crisis indicated the relative frailty of the Western world due to its dependence on the supply of that product. If Antarctica could provide an alternative source, that frailty could be overcome.

To these elements—the pressure imposed by the IBEA planned Antarctic expedition and the rising interest in the prospects of mineral exploration and exploitation in Antarctica—a third factor could be added to make the Brazil's incursion in the Antarctic politics possible: the election of Ernesto Geisel as President of Brazil.

2.3 Brazil's Step into Antarctic Politics

Coming from the military branch and having actively participated on the *coup d'état* of 1964, along with many other important political events, Ernesto Beckmann Geisel was President of the State-owned oil company Petrobras when he was selected as successor to the Presidency under the military regime. The choice of Geisel was not fortuitous as Brazil was dealing with a slowdown of its previously impressive economic performance as a consequence of the impact of the economic oil crisis on Brazil's economy. The effects of the increase of the international price of

[98] The Antarctic Treaty and other key documents can be consulted at the Antarctic Treaty Secretariat website: www.ats.aq.

crude oil was of particular relevance to Brazil's incipient industry as the country's energy system was highly dependent on crude oil imports.[99]

In addition to the energy crisis, Geisel's Presidency had to deal with many other challenges both domestically and internationally. Domestically, the tensions generated by the brutal military regime were reaching a critical point, reinforced by international criticism of its authoritarian and repressive character. Internationally, the crisis with Argentina and misunderstandings with the US led to a re-reading of Brazil's international alliances. These added factors made Geisel search for new directions in Brazil's politics. To assist him in that task, Geisel called the prominent geopolitician Golbery de Couto e Silva to serve as Chief of Staff of the Presidency. The appointment of Silva caused unease in Argentina, where his geopolitical ideas were well known.

Domestically, Geisel started what was known as the '*apertura*' (opening) process, lessening the political restraints and derogating the infamous 'Institutional Act 5' that had suppressed the constitutional guarantees.[100] While this policy did not please many of the military hardliners, it allowed Geisel to avoid an escalation of popular resistance to his government. On the other hand, the 'opening' did not mean the end of the military repression, and Geisel's government combined that increased political freedom with a less numerous but equally violent detentions, torture and summary executions of what was considered as 'subversive' elements.

Internationally, Geisel adopted what was called a 'responsible pragmatism,'[101] basically translated to a wider scheme of alliances and the

[99] It could even be argued that a great deal of its industrial development on the previous years was allowed by the availability of low-priced oil.

[100] The infame *Ato Institucional Nro. 5* (Institutional Act N° 5) was the fifth Decree issued by the Military Dictatorship (signed by Costa e Silva on 13th December 1968) aroused from the *coup d'état* of 1964, which completed the assertion of direct control of the country's political life by the military and suspended the constitutional guarantees.

[101] The so-called *Pragmatismo Ecumênico e Responsável* (Ecumenical Responsible Pragmatism) looked to take advantage of what was seen as a more open international system in pursuance of a more independent international policy. It was based on a political realist approach but without abandoning the traditional principles of Brazil's diplomacy. For details on the origins and characteristics of the Responsible Pragmatism, see: Spektor (2004).

loosening of the ideological barriers. In practical terms, Brazil recognized the People's Republic of China and Angolan independence,[102] and initiated negotiations with West Germany for the transference of nuclear-power-generation technology. In terms of its relations with Latin-America, Geisel tried to balance its relationship with Argentina at a time when tensions were rapidly rising.

In the wake of the conflict over the Itaipu dam project, Geisel looked for a rapprochement with Argentina that could lead to a possible solution to the dispute without renouncing to the needs of developing an alternative source of the generation of electricity. With the hydroelectric project as the spine of the alternative generation options, Geisel prioritized the former without renouncing to other possibilities. That was consistent with the choice of not antagonizing Argentina in the issue of Antarctica, avoiding aggravation of the existing tensions, nonetheless, not renouncing to gradually entering into scene.

As President of Petrobras, Geisel invested heavily in the improvement of the refinery infrastructure but, for undetermined reasons, he did not consider it timely to invest in prospection. That changed in 1975, during his Presidency, coincidently with the change on the position regarding Antarctica. As the documentation available suggests, since the end of 1971, the EMFA requested information regarding scientific activities and bases operating in Antarctica, for which it instructed the Brazilian embassies in AT member countries to gather such information.[103]

[102] Both cases reveal a tipping point in the Brazilian attitudes framed by the Cold War. The recognition of communist China and the Republic of Angola—the largest and richest Portuguese colony in Africa—marked a tendency to broaden the range of international relations and alliances, even with countries under communist rule.

[103] See documentation present in: Antártida. – Dossiê (BR DFANBSB Z4.REX.IPS.26), Arquivo Nacional, Brazil. It includes information regarding Antarctic activities of several countries and press articles from a diverse number of sources. Since the end of 1972 was predominant the press reports from Argentina with a negative view about the Brazilian attitude. The documentation includes reports from Azeredo da Silveira, who was at that time Ambassador in Buenos Aires and eventually Minister of Foreign Affairs under Geisel's Presidency, in which he informed about the negative reactions to the possibility of Brazilian involvement in Antarctica (see, for example: Expedicao Brasileira A Antartida. Brasemb Baires, 9 February 1973; Antártida. Comentarios da Imprensa Argentina. Brasemb Baires, 30 March 1973 and Antartida. Politica Argentina; Brasemb Baires, 14 August 1973; ibid.).

In May 1974, the Brazilian Embassy in Washington revealed an article by Deborah Shapley, published in *Science* magazine[104] which reported that the US National Security Council was considering proposing a review to the AT, as a consequence of the possibility of mineral exploitation in the continent, including oil.[105] By that time, efforts to obtain Congress approval to provide logistical support to the IBEA expedition confronted important obstacles in the position adopted by Itamaraty. As expressed by the Brazilian Ambassador in Buenos Aires, João Batista Pinheiro, to the International Relations Commission of the Congress, the Foreign Ministry considered it untimely to take any initiative in that direction.[106]

That same month, the Ministry of Foreign Affairs raised the question of accession to the AT with the Presidency, something that was probably fuelled by both the alleged prospects of revision of the treaty and the planned expedition of the IBEA.[107] The report stressed the long-term interest of Brazil in the exploitation of Antarctic natural resources and the short- and medium-term interests of national security and scientific research.[108] From all claims advanced up to the time, the report considered those of Argentina and Chile the most substantial. Regarding the AT to be an established fact, it considered more convenient to work within such international framework than to oppose to it. Under these considerations, and looking for the support of its continental neighbours, the recommendation was to avoid contradicting the other south American countries' interests in the region, defending instead the right of those countries more 'bonded' to the continent to participate. The report also suggested to access to the Treaty, with or without a declaration restating Brazil's position on the matter; to participate in Antarctic activities, by itself or in collaboration with other Treaty members; and to wait to a more appropriate moment for advancing the arguments of Brazilian rights in Antarctica.

Despite several public declarations revealing important strains between Buenos Aires and Brasília regarding Antarctica on the previous year, in

[104] Shapley (1974).

[105] Castro (1974).

[106] Melhores relações Brasil-Argentina (1974).

[107] The reference is to the *Exposição de Motivos n° DAM/1/G182/930(B46) (D)* of the Ministry of Foreign Affairs, dated on 28 May 1974 (see: Abreu, 1974b).

[108] Abreu (1974a, pp. 6–7). See also Ferreira (2009, p. 127).

July 1974 there were signs that both Argentina and Chile were in favour of engaging with Brazil in some form of international cooperation on Antarctic scientific activities, if that could prove of use in challenging the position of the Northern Hemisphere powers.[109] Pinochet's military coup of the Socialist government of Salvador Allende in Chile facilitated the rapprochement between Santiago and Brasília. In August 1974, at a private dinner between the Brazilian Ambassador in Santiago and the Chilean Ministry of Foreign Affairs, the latter expressed his opinion that Chile's military government would eventually support a Brazilian claim if the Treaty was abandoned in 1991.[110] Furthermore, it offered the Ambassador the opportunity to join the next Chilean expedition to Antarctica and to assist the Brazilian government with the Chilean Navy's knowledge and expertise in the region. Although the Brazilian Ambassador promptly accepted the Chilean invitation, he was later instructed to reject the offer as it could be interpreted as an official recognition of the Chilean position regarding Antarctica.[111]

In September, Itamaraty insisted on its position sending an additional report to the Presidency. The report stated the alleged support of Argentina and Chile to Brazil's participation in activities in the southern regions, noted the weakened position of 'territorialist' countries within the 'polar club' and stressed the convenience of joining the Treaty in order to be able to participate in its revision in 1991.[112] A few days later, the Secretary-General of the National Security Council of Brazil considered the positions forwarded by Itamaraty on its report of May and instructed that its own report be submitted to the President.[113] After summarizing the content of the Itamaraty's report, it concluded that Antarctica had multiple aspects linked with Brazil's national security, that Brazil had interests in Antarctica rather than rights, that Brazil needed

[109] Pinheiro (1974).

[110] Presenca do Brasil em Arica/Antartica (1974) and Relações Brasil-Chile, Antártida (1974).

[111] Viagem À Antartida (1975a, 1975b).

[112] Silveira (1974).

This was motivated by a common misreading of the Treaty, which was seen as establishing an 'expiration' date thirty years after it enters into force when, actually, it only made more flexible the process for requiring its revision. For some details on Itamaraty's report, see: Ferreira (2009, p. 127).

[113] The reference is to the *Exposição de Motivos N°61/74* (Abreu, 1974a).

to adopt a pragmatic attitude taking into consideration long-term objectives, and suggested that the accession to the AT, as recommended by Itamaraty, stood as the better course of action. In doing so, the report recommended to establish safeguards to Brazil's interest in the continent and to obtain the maximum freedom of manoeuvre possible. As a final remark, it justified the assignment of Antarctic decisions within the scope of the National Security Council, due to its long-lasting and long-reaching effects. Despite Itamaraty and the National Security Council's efforts, the brief Presidential response to the suggestion to accede to the Treaty was to postpone such a decision until a more fitting time.[114]

Even when Itamaraty was not considering the possibility of advancing a formal claim to Antarctica, it is notable that, in January 1975, just a few months before Brazil's accession to the AT, they initiated consultations with the Brazilian Historical and Geographical Institute of Rio de Janeiro to confirm the veracity of the versions that the Belgian Expedition, under the command of Adrien de Gerlache, had flown the Brazilian flag in the white continent. The Institute confirmed the account of Gerlache that, on 28 October 1899, the Brazilian flag was hoisted on the main mast of the expedition ship *Belgica*, in gratitude for the collaboration given in their stop in Rio de Janeiro.[115] The issue seems not to have generated any further action, but it was later recalled each time that Brazil's antecedents in Antarctica were considered.

The next month, during the meetings of the Special Committee for Review of the Inter-American System, the Brazilian delegation opposed any mention of the 'American Nations' claims in the Antarctic region and the inclusion of a map describing the extension of such claims. Brazil's participation restated the position established on the confidential note delivered in 1958 to the US government, in which they had protested against their exclusion from the Antarctic conference.

On 2 May 1975, the Ministry of Foreign Affairs submitted a new report to the Presidency, insisting on the accession to the AT.[116] Once more, the report stressed the political relevance of the AT, considered the

[114] Abreu (1974a, 1974b).

[115] There are several versions within the Brazilian literature that describes the event as the first time that a Brazilian flag was hoisted in Antarctic territory. Actually, following Gerlache's account, the flag was hoisted in the main mast of the ship and not 'in land.'

[116] Silveira (1975).

first and solely juridical instrument for Antarctica. It stressed its importance in avoiding the extension of Cold War tensions to the southern region and considered it a precursor for possible international understanding in other critical issues, such as outer space and deep-sea bed. Considering the Antarctic regime as a form of limited internationalization, the report highlighted the freezing of claims and the inspection mechanism provided by the Treaty; the promotion of scientific research and international collaboration to that end; and the prohibition of nuclear activities, including the disposal of radioactive material. On the other hand, it warned about the 'expiration' of the Treaty in 1991[117] and the omission of provisions regarding the exploitation of Antarctic natural resources, considering that new economic prospects were being developing in what regarded oil and natural gas. Furthermore, the report pondered the historical relevance of raising the Brazilian flag in Antarctica by Gerlache, the collaboration given to Charcot, the interest demonstrated by prominent figures in non-official channels, the diverse number of studies developed within government circles and the initiative taken by the Engineering Club of Rio de Janeiro as relevant record of Brazil interest in the region.

In addition, the report analysed the possibilities open to accession to the Treaty. Such action, taken by that time by four countries besides the original 12 signatories, was considered a prerequisite to undertake activities in the continent, in order to conduct 'substantial scientific activity' and be allowed to participate in the decision-making process. The requirement of scientific activity as a condition for participating in the decision-making process was considered discriminatory against the acceding countries, as well as the provision of Article IV freezing rights and claims.[118] However, such discriminatory effects were seen as being toned-down by the course of events and the possibility of a de jure and de facto presence of Brazil in Antarctica. Regarding scientific activities, the report highlighted the need of Brazil to obtain Argentina's or Chile's support in order to conduct activities in the region, which would force it to obey the Treaty provisions.[119] Regarding rights and claims, the report

[117] As seen, this was a misreading of the Treaty.

[118] Itamaraty's views on that regard was that such condition was not imposed to the original signatories.

[119] Argentina and Chile had jointly declared that any activity undertaken in Antarctica should follow the Antarctic Treaty provisions. As almost every expedition that headed to

considered that, since only seven countries had advanced claims at the time of signature of the Treaty, and that there were already nine countries that had not advanced any claim, the position of the 'territorialist' countries (the claimant countries) had been weakened by the time the Treaty was enforced; and considered that Article IV actually had operated more in favouring the internationalization of the continent than to assuring the rights of claiming countries.

The conclusions of Itamaraty's report recommended the accession to the Treaty, based on many of the aspects advanced on the report of May 1974 and reiterating the short- and medium-term interests of national security and long-term interest of participating on the exploitation of the continent natural resources. This time, the importance of conducting scientific studies of a general character but also related to meteorology and other aspects directly linked with effects over the Brazilian territory was explicitly raised.[120] Additionally, it was considered that, as the reservation of rights on the Brazilian note of 30 July 1958 preceded the signature of the AT, the accession to the Treaty would not imply any renunciation. In any case, that position could be restated by issuing a declaration after Brazil's accession to the Treaty at a timely opportunity. In terms of the potential effects, the accession of Brazil to the AT was expected to generate interest in other countries such as Ecuador, Peru and Uruguay, forcing a wider multi-lateralization of scientific research and favouring regional collaboration. However, it was also stated that, if the Treaty was to expire and the 'frontage' criteria were to prevail, Brazil should make its 'incontestable rights,' based on its geographical position, to prevail.

Considering the discussions about exploitation of Antarctic natural resources on the upcoming consultative meeting of the AT, the report highlighted the interest that the issue was getting, including the findings of potential oil reserves on the seabed by the US *Glomar Challenger* expedition. The UK proposal of creating of some form of supranational organism to administrate the exploitation of natural resources was

the Antarctic Peninsula would have to stop in one of those countries' ports, they had an opportunity to enforce such attitude.

Where it is not clarified in the report the justification of the need to cooperate with Argentina and/or Chile, it could have obeyed the need of calling into Argentine or Chilean ports. It is unclear why they did not considered collaboration with the British. Nonetheless, in that case, their conclusion about the pre-requisite of obeying the Treaty would not have changed neither.

[120] Silveira (1975).

noted, considering that inclusion in such discussions was opportune and necessary to guarantee Brazil's interests.

The report was submitted by the President to the National Security Council for its consideration. The Council consulted the Minister of Justice, Armando Falcão, who, on 14 May, agreed with the reflections of the Itamaraty's report regarding the convenience to adhere to the international agreement.[121] However, adopting a more realistic view, Falcão warned that such a move could be interpreted by the other members as anything other than a materialization of Brazil's interest in the Antarctic, and that some opposition could arise. Added to that, the Minister expressed doubts about the possibility of participating in the next consultative meeting as, in his understanding, the unilateral accession to the Treaty would not serve the condition established in the Treaty of substantial scientific activity. However, instead of preventing Brazil from being incorporated to the Treaty, the Minister suggested that those considerations indicated that the accession should have been done earlier, and that such an action remained imperious.

On 16 May 1975, the National Security Council reported to the President that, given the previous year's decision and in consideration of the new elements presented by the Ministry of Foreign Affairs, they concluded that the accession to the AT was in furtherance of national security. Furthermore, if such course was to be adopted, it would be opportune to make efforts to accelerate Congress approval.[122] That same day, by a handwritten note about the Council's report that read '*I Agree, on 16 May 75*,'[123] President Geisel finally endorsed the proposal and, with it, Brazil acceded to the AT.

Also on that day, the Brazilian Ambassador in Washington, João Augusto de Araujo Castro, delivered to the Treaties Division of the US Department of State a note through which Brazil acceded to the AT.[124] In the press release delivered to make the decision public, the government recognized the AT as the solely legitimate legal instrument to the Antarctic continent and to which all activities in the region should be subject. At the same time, it recognized the positive values of pacific

[121] Falcão (1975).

[122] Abreu (1975a).

[123] '*Aprovo. Em 16 maio 75*' (Abreu, 1975a, 1975b).

[124] Adesão do Brasil ao tratado da Antartida (1975a, 1975b).

use, the system of inspections, the freedom of scientific research and the prohibition of nuclear explosion and radioactive material disposal. It also acknowledged the extreme importance of all scientific research that was being undertaken on the continent. In considering Brazil's position, the press release emphatically stated that the principle of equality on the decision-making process of the regime should prevail; that the possession of the most extensive South Atlantic coast results in substantial and direct interest in the continent; and that the importance of the region resulted in its inclusion of the Inter-American Treaty of Reciprocal Assistance, to which Brazil is co-responsible.

Almost immediately, the Brazilian legation in Washington started to press to be allowed to participate on the upcoming Consultative Meeting. It was informed that, as there was no permanent secretary of such instrument, the consultation was to be done with the organizing country, in that instance, the Norwegian government. In spite of the original intention of requiring an observing status, Brazil ended up insisting on their interpretation that the country already held substantial tradition and interest in Antarctica and that it considered itself as holding the right to participate as a full member of the consultative meeting. In a manoeuvre to look for an alliance with Brazil, in June 1975, the British Foreign Office approached the Brazilian government offering the facilities at the Scott Polar Research Institute at Cambridge and to take part in the upcoming British Expeditions to Antarctica.

On 16 June 1975, after the proposal for accession to the AT had been approved by the Constitutional and Justice Commission, the issue was debated in Congress. During the debate, the usual arguments and basis of Brazilian rights in Antarctica were stated, adding to the declared interest in the southern sea and its resources—including the presence of the blue whale, of 125 tons, the meat of which could feed 280,000 people!—and the strategic importance due to its meteorological influence over Brazilian territory.[125] The Congressional approval was uneventful, and the speakers expressed mixed opinions that combined the importance of the AT and the alleged Brazilian rights to take advantage of the resources present on the continent, including a recall of the principles defended by the so-called frontage theory, forwarded by Terezinha de Castro, Menezes and

[125] Diario do Congresso (1975, pp. 4306–4310A).

The estimates of the number of people that a whale can feed correspond to the declaration of Representative Hugo Napoleão.

others. A month-and-a-half after the Presidential decision, on 29 June 1975, the Brazilian Congress issued the legislative decree N°56, granting its consent to Brazil's accession to the Treaty.

On 11 July 1975, President Geisel issued the National Decree 75,963, through which the AT was promulgated. Brazil was finally part of the Treaty, but there was still a long way to go to demonstrate that its commitment with the white continent was more than just words.

2.4 Conclusions

When compared with Argentina and Chile, Brazil could be considered as a latecomer onto the Antarctic scene. However, that appraisal needs to be qualified. It is true that Brazil had not demonstrated any active interest in Antarctica up until the second half of the twentieth century and that, besides some isolated manifestations of interests, it took until the last quarter of that century to take a step into Antarctic politics. Nonetheless, Brazil was the first Latin-American country to accede to the Treaty after Argentina and Chile, and the only Latin-American country besides those two to elaborate a specific position regarding territorial claims that became notorious.

Despite a sector of the Brazilian elite expressed interest in the country's active participation in Antarctica, including the possibility of advancing a territorial claim and the dispatch of a scientific expedition, several factors acted against an earlier involvement. Limitations imposed by a lack of interest, domestic needs, structural conditions and the priority given to other international issues determined that the effective adoption of a position regarding Antarctica was continuously deferred.

Up until after the World War II, Brazil's interests were directed to the hinterland. But as the country's economic matrix, society and international stand changed, it became more interested in the overseas projection of power. The international context of the Cold War led the Brazilian geopoliticians to think the place of the country within the global order. However, Antarctica played a rather marginal role in that order and, with the exception of Terezinha de Castro, little importance was given to the Antarctic region.

Also, by the middle of the twentieth century, the Brazilian scientific system was at its infancy. The higher educational system inherited from the Colony hindered the advancement of a national system of research and development. Many key disciplines involved in Antarctic research were underdeveloped, and there was no constituted community of scientists interested in Antarctic research. Only thanks to the rapid development of the scientific system in Brazil and the rising interest in part of the intellectuality, a portion of the academic elites started to push for an Antarctic scientific endeavour in the 1970s.

In terms of Brazil's foreign policy, the pragmatism of the Brazilian Ministry of Foreign Affairs led a reluctance to take a position, looking to keep its options open while also avoiding the rebuttal of Argentina and Chile. That was reinforced by the Itaipu dam dispute, as it prioritized the need to maintain good relations with Argentina in front of the more immediate interest in the hydroelectrical project. In sustaining that policy, the autonomy of Itamaraty was essential, as it allowed the government to resist pressures from influential sectors of the Brazilian society, including the ESG, the Rio de Janeiro Engineering Club, etc.

Another factor that must have played a role in deferring the decision was the complex system of decision-making installed after the military coup of 1964. The involvement of the National Security Council, Itamaraty, the Presidency and the Ministry of Justice stretched the decision-making process, involving a variety of actors and making it more difficult to determine a specific decision when its determinants were not very clear. On the other hand, Congressional participation in the decision was token, as shown by the lack of power in favouring the IBEA expedition and any real participation in the decision of acceding to the Treaty.

References

Anuário do Instituto Rio-Branco 1956–1957. (n.d.).

Azambuja, P. (2005). *O sonho do aurora austral: como o Brasil chegou à Antártida*. Magna Quies.

Barr, S., & Lüdecke, C. (Eds.). (2010). *The history of the International Polar Years (IPYs)*. Springer-Verlag.

Beck, P. J. (1985). Preparatory Meetings for the Antarctic Treaty 1958–1959. *Polar Record*, *22*(141), 653–664.

Borges, D. R. S. (1959). *Um brasileiro na Antártida; crônicas de uma viagem ao redor do mundo*. Sociedade Geográfica Brasileira.

Bradford Burns, E. (1966). *The unwritten alliance: Rio Branco and Brazilian-American relations*. Columbia University Press.

Cardone, I. J. (2019). *A continent for peace and science: Antarctic science and international politics from the 6th International Geographical Congress to *the Antarctic Treaty (1895–1959)*. University of São Paulo—King's College London. https://doi.org/10.11606/T.101.2019.tde-16082019-164433

Cardone, I. J. (2020). Shaping an Antarctic identity in Argentina and Chile. *Defence Strategic Communications, 8*, 53–88. https://doi.org/10.30966/2018.RIGA.8.2

Cardone, I. J., & Fontana, P. G. (2019). Latin-American contributions to the creation of the Antarctic regime. *The Polar Journal, 9*(2), 300–323. https://doi.org/10.1080/2154896X.2019.1685174

Carvalho, D. de, & Castro, T. de. (1956). A Questão Antártica. *Boletim Geográfico, XIV*(135), 502–506.

Carvalho, D. de, & Castro, T. de. (1960). *Atlas de Relações Internacionais*. IBGE.

Castro, T. de. (1959). Antártica o Assunto do Momento. *Boletim Geográfico, XVII*(150), 238–245.

Castro, T. de. (1976). *Rumo à Antártica*. Freitas Bastos.

Cervo, A. L., & Bueno, C. (2012). *História da política exterior do Brasil* [4. ed.]. Brasília: Editora Universidade de Brasília.

Chaturvedi, S. (2013). Rise and decline of Antarctica in Nehru's geopolitical vision: Challenges and opportunities of the 1950s. *The Polar Journal, 3*(2), 301–315. https://doi.org/10.1080/2154896X.2013.868087

Coffman, S. E., Sampson, C. S., & Glennon, J. P. (Eds.). (1991). *Papers relating to the foreign relations of the United States, 1958–60* (Vol. II, e-book). United States Government Printing Office.

Cook, F. A. (1909). *Through the first Antarctic night*. Doubladay, Page & Company.

de Oliveira, C. G. F., & Pereira, W. E. N. (2016). Coronelismo e direito eleitoral na Velha República. *Revista Interfase, 13*(2), 65–77.

Dias De Oliveira, N. (2010). Os Primórdios da Doutrina de Segurança Nacional: A Escola Superior de Guerra. *História, 29*(2), 135–157. http://www.scielo.br/pdf/his/v29n2/v29n2a08

Fajardo, J. M. C. (2004). *Acordo Tripartite Itaipu-Corpus: ponto de inflexão entre a disputa geopolítica e a política de cooperação*. Universidade Federal do Rio Grande do Sul. http://www.lume.ufrgs.br/handle/10183/6148

Ferreira, F. R. G. (2009). *O Sistema do Tratado da Antártica: evolução do regime e seu impacto na política externa brasileira*. Fundação Alexandre de Gusmão.

Fontana, P. (2014). *La Pugna Antártica: el conflicto por el sexto continente*. Guazuvirá Ediciones.

Foresta, R. A. (1992). Amazonia and the politics of geopolitics. *Geographical Review, 82*(2), 128–142.

Hall, H. R. (1994). *International regime formation and leadership: The origins of the Antarctic treaty*. University of Tasmania. Retrieved from http://eprints.utas.edu.au/11401/2/Hall_1994.pdf

Hepple, L. W. (1986). Geopolitics, generals and the state in Brazil. *Political Geography Quarterly*. https://doi.org/10.1016/0260-9827(86)90059-5

Howkins, A. (2006). Icy relations: The emergence of South American Antarctica during the Second World War. *Polar Record, 42*(2), 153–165. https://doi.org/10.1017/S0032247406005274

Howkins, A. (2008a). Defending polar empire: Opposition to India's proposal to raise the 'Antarctic Question' at the United Nations in 1956. *Polar Record, 44*(01), 35–44. https://doi.org/10.1017/S0032247407006766

Howkins, A. (2008b). *Frozen empires: A history of the Antarctic sovereignty dispute between Britain, Argentina, and Chile, 1939–1959*. The University of Texas at Austin.

Mattos, L. F. de. (2015). *O Brasil e a Adesão ao Tratado da Antártica: uma análise de política externa no governo Geisel*. Universidade Federal Fluminense.

Menezes, E. C. de. (1982). *Antartida e os Desafios do Futuro*. Capemi Editora.

Musto, R. A. (2017). Cold calculations: The United States and the creation of Antarctica's atom-free zone. *Diplomatic History, 42*(4), 640–668. https://doi.org/10.1093/dh/dhx045

Nickson, R. A. (1982). The Itaipú Hydro-Electric Project: The Paraguayan perspective. *Bulletin of Latin American Research, 2*(1), 1–20.

Peterson, M. J. (1988). *Managing the frozen south: The creation and evolution of the Antarctic treaty system*. University of California Press.

Pinheiro, L. (2010). *Política Externa Brasileira (1889–2002)* (2nd ed.). Jorge Zahar Editor Ltda.

Pinto Coelho, A. (1983). *Nos confins dos três mares... a Antártida* (Ed. Rev. e). Biblioteca do Exército Editora.

Ricupero, R. (2018). José Maria da Silva Paranhos Júnior (Barão do Rio Branco): A Fundação da Política Exterior da República. In J. V. de S. Pimentel (Ed.), *Pensamento Diplomático Brasileiro: Formuladores e Agentes da Política Externa (1750–1964)* (Vol. II, pp. 405–438). Fundação Alexandre de Gusmão.

Riso Patrón, L. (1908). La Antártida Americana. *Anales de La Universidad de Chile, 122*, 243–265. https://doi.org/10.5354/0717-8883.2012.24573

Rodrigues, V. de J. (1998). *O Sistema do Tratado da Antártica: Implicações para o Direito Internacional*. Universidade de São Paulo.

Rosenbaum, H. J. (1973). Argentine-Brazilian relations: A critical juncture. *The World Today, 29*(12), 537–542. https://www.jstor.org/stable/40394681

Schwartzman, S. (1978). Struggling to be born: The scientific community in Brazil. *Minerva, 16*(4), 545–580. https://doi.org/10.1007/BF01100332

Shapley, D. (1974). Antarctica: World hunger for oil spurs security council review. *Science, 184*(4138), 776–780.

Spektor, M. (2004). Origens e direção do Pragmatismo Ecumênico e Responsável (1974–1979). *Revista Brasileira de Política Internacional, 47*(2), 191–222. https://doi.org/10.1590/s0034-73292004000200007

Sullivan, W. (1962). *Assault on the unknown. The international geophysical year [With plates]*. Hodder & Stoughton.
Swinney, G. N. (2001). Some new perspectives on the life of William Spiers Bruce (1867–1921), with a preliminary catalogue of the Bruce collection of manuscripts in the University of Edinburgh. *Archives of Natural History, 28*(3), 285–311.
Swinney, G. N. (2007). The Scottish national Antarctic expedition (1902–04) and the founding of Base Orcadas. *Scottish Geographical Journal, 123*(1), 48–67. https://doi.org/10.1080/00369220718737283
Villela, R. J. (1962). Resultados geográficos e meteorológicos da expedição da "Operação Deep Freeze 1961" ao Mar de Bellingshausen (Antártida). *Ciência e Cultura, 14*(4), 267–268.
Wiltgen, J. A. (1977). Instituto Brasileño de Estudios Antárticos: Objetivos y Políticas. In F. O. Vicuña & A. S. Araya (Eds.), *El Desarrollo de la Antártica* (pp. 324–326). Editorial Universitaria - Instituto de Estudios Internacionales de la Universidad de Chile.

Archival Sources

Abreu, H. A. (1974a, September 8). [Report on Brazil and Antarctica]. Arquivo Antônio Azeredo da Silveira - Exposição de Motivos 061/74 (AAS mre ag 1974.01.16/DOC I-2A1), CPDOC-FGV, Brazil.
Abreu, H. A. (1974b, September 11). [Exposição de Motivos N° DAM/1/G182/930(B46) (D)]. Arquivo Antônio Azeredo da Silveira – Aviso N°287/74 (AAS mre ag 1974.01.16/DOC I-2), CPDOC-FGV, Brazil.
Abreu, H. A. (1975a, May 16). [Exposição de Motivos DAM-I/124/210(B29)] Divisão de Segurança e Informações do Ministério da Justiça - BR RJANRIO TT, pp. 45–47 - Exposição de Motivos N035/75 (BR/RJANRIO/TT/0/JUS/PRO/0538), Arquivo Nacional, Brazil.
Abreu, H. A. (1975b, May 16). [Exposição de Motivos DAM-I/124/210(B29)] Divisão de Segurança e Informações do Ministério da Justiça - BR RJANRIO TT, p. 44 – Aviso N192/75 (BR/RJANRIO/TT/0/JUS/PRO/0538), Arquivo Nacional, Brazil.
Adesão do Brasil ao Tratado da Antártida. Nota ao Departamento de Estado. (1975a, May 16). [Notification of Brazil's accession to the Antarctic Treaty]. Antártida. – Dossiê, p. 759 (BR DFANBSB Z4.REX.IPS.26), Arquivo Nacional, Brazil.
Adesão do Brasil ao Tratado da Antártida. Nota ao Departamento de Estado. (1975b, May 16). [Notification of Brazil's accession to the Antarctic Treaty]. Antártida. – Dossiê, p. 758 – G/SG/DAA (BR DFANBSB Z4.REX.IPS.26), Arquivo Nacional, Brazil.

Castro, J. A. de A. (1974, May 16). [Report about Deborah Shapley in Science Magazine]. Antártida. – Dossiê. p. 821 - Antartida. Exploracao de Recursos Minerais... (BR DFANBSB Z4.REX.IPS.26), Arquivo Nacional, Brazil.

Copias de las Actas Firmadas en Santiago de Chile.... (1941, June 4). [Proceedings of the Negotiations Between Argentina and Chile in 1941]. Exp N° 16 – Soberanía en Tierras Antárticas. (AH/0003/4), Archivo Histórico de la Cancillería, Buenos Aires, Argentina.

Estado-Maior das Forças Armadas. (1970, August). [Report of domestic and international context] Fundo: Estado-Maior das Forças Armadas – Documentos Sigilosos Expedidos, pp. 2–16 - Boletim Mensal de Informações N°8/70 (BR/DFANBSB/2M.0.0.5), Arquivo Nacional, Brazil.

Falcão, A. (1975, May 14). [Letter to Abreu] Divisão de Segurança e Informações do Ministério da Justiça - BR RJANRIO TT, pp. 40–43 - AV.G/000417-B (BR/RJANRIO/TT/0/JUS/PRO/0538), Arquivo Nacional, Brazil.

Maury, M. F. (1861, April 10). [Letter to His Excellency, The Counsellor, Senhor M. M. Lisboa, Envoy Extraordinary & Minister Plenipotentiary of Brazil] Records of the U.S. Naval Observatory (RG 78/Entry I/Vol 17), US National Archives Building, Washington, DC, US.

Melhores relações Brasil-Argentina. (1974, May 24). [Clipping from Folha de São Paulo newspaper]. Antártida. – Dossiê, p. 819 (BR DFANBSB Z4.REX.IPS.26), Arquivo Nacional, Brazil.

Peixoto, E. A. (1958, November) [Letter from the Brazilian Ambassador in the US to the Brazilian Secretary of State] Arquivo Ernani do Amaral Peixoto – Anártica, Uso Pacífico (EAP emb 1956.01.03), CPDOC-FGV, Brazil.

Pinheiro. (1974, July 19). [Report on conversations with Argentine Adm. Tagliatti] Antártida. – Dossiê, p. 811 – Brasemb Baires, Grupo Internacional de Coordenação do Oceano Antártico (BR DFANBSB Z4.REX.IPS.26), Arquivo Nacional, Brazil.

Presenca do Brasil em Arica/Antartica. (1974, September 5). [Report on conversations with Chilean Minister of Foreign Affairs] Antártida. – Dossiê, p. 810 - Brasemb Santiago, G/DAA/DAM-I/ (BR DFANBSB Z4.REX.IPS.26), Arquivo Nacional, Brazil.

Relações Brasil-Chile, Antártida. (1974, September 5). [Report on conversations with Chilean Minister of Foreign Affairs] Antártida. – Dossiê, pp. 805–807 - Brasemb Baires, DAM-I/ (BR DFANBSB Z4.REX.IPS.26), Arquivo Nacional, Brazil.

Secretaria de Estado das Relações Exteriores to American Embassy. (1958, July 30). [Letter regarding negotiations for an Antarctic regime and Brazil's exclusion] Divisão de Segurança e Informações do Ministério da Justiça - BR RJANRIO TT, pp. 38–39 (BR/RJANRIO/TT/0/JUS/PRO/0538), Arquivo Nacional, Brazil.

Secretaria Geral do Conselho de Segurança Nacional. (1974, July 30). [Report to the General Secretary]. Arquivo Antônio Azeredo da Silveira - Informação N067/la.SC/74 (AAS mre ag 1974.01.16/DOC I-2A2), CPDOC-FGV, Brazil.

Silveira, A. F. A. (1974, September 5). [Accession to the Antarctic Treaty]. Arquivo Antônio Azeredo da Silveira – O Tratado da Antártida, Consequências da Adesão (AAS mre d 1974.03.26/DOC III-48), CPDOC-FGV, Brazil.

Silveira, A. F. A. (1975, May 2). [Accession to the Antarctic Treaty]. Divisão de Segurança e Informações do Ministério da Justiça - BR RJANRIO TT, pp. 4–26 - DAM I/124/210(B29) (BR/RJANRIO/TT/0/JUS/PRO/0538), Arquivo Nacional, Brazil.

Viagem à Antartida. (1975a, January 6). [Offer of Chile to join the next Antarctic expedition] Antártida. – Dossiê, pp. 792–793 - Brasemb Santiago, G/SG/DAM-I (BR DFANBSB Z4.REX.IPS.26), Arquivo Nacional, Brazil.

Viagem à Antartida. (1975b, January 8). [Offer of Chile to join the next Antarctic expedition] Antártida. – Dossiê, p. 791 - Brasemb Santiago, G/DAM-I (BR DFANBSB Z4.REX.IPS.26), Arquivo Nacional, Brazil.

Newspapers Articles and Other Sources

Brasil Precissa Descobrir a Antártida. (1971, March 14). *O Estado de São Paulo* (São Paulo, Brazil), 128.

Decreto N° 75,963. (1975, July 11). Brazil.

Decreto Legislativo N°56. (1975, June 29). Brazil.

Defrontação não é a melhor rota para o Brasil chegar à Antártida, diz cientista. (1973, February 28). *Jornal do Brasil* (Rio de Janeiro, Brazil), Issue 312, 7.

Diario do Congresso Nacional. (1975 June 17). Seção I. Vol 30(063). Brasília, 4306–4310. http://imagem.camara.gov.br/Imagem/d/pdf/DCD 17JUN1975.pdf. Accessed on 5 September 2019.

Embarcou para o Pólo Sul o professor Durval Borges. (1958, February 12). *Correio da Manhã* (Rio de Janeiro, Brazil), Issue 19903.

Expedição brasileira à Antártida talvez não passe do sonho de cientistas. (1973, February 21). *Jornal do Brasil* (Rio de Janeiro, Brazil), Issue 305, 19.

Foi para a Antártida. (1958, January 14). *Correio da Manhã* (Rio de Janeiro, Brazil), Issue 19878.

Guerra Fria entra em sua fase gelada. (1957, January 21). *Manchete* (Rio de Janeiro, Brazil), Issue 296, 24–27.

Menezes. (1971, June 5). A Antártida também é nossa. *Manchete* (Rio de Janeiro, Brazil), Issue 998, 70–80.

Meteorologista brasileiro que foi ao polo sul conta a viagem. (1961, December 31). *O Estado de São Paulo* (São Paulo, Brazil), 17.

Ribeiro, J. (1955, April 30). Um pedaço da Antártida pertence ao Brasil. *Revista da Semana* (Rio de Janeiro, Brazil), Issue 18, 44–47.
Visconti, A (1958, April 19). O Brasil também quer uma fatia do pólo sul. *Manchete* (Rio de Janeiro), Issue 313, 30–33.

CHAPTER 3

When the Tropic Met the Pole

3.1 The Search for Consultative Status

The efforts of Brazil to be acknowledged as a member with consultative status—that is with the right to participate in the discussion and decision-making process of the Antarctic regime—based on its tradition and interests were fruitless. As Minister Falcão had anticipated, the condition established in Article IX of demonstrating substantive scientific research was not reached by Brazil. Up until then, the country could only display a few papers elaborated by Brazilian scientists who participated as guests in expeditions organized by other nations.[1] For the Antarctic Treaty (AT) consultative parties, this fell short of demonstrating a substantial commitment to Antarctica and could open the door to other countries by the mere statement of their interest.

As the Brazilian government realized it would need an active involvement in Antarctica to participate on the decision-making of the regime, it started to study which institutional design would better suit its interests. Since June 1975, the Brazilian government had sent instructions to its

[1] A search conducted on 13 September 2019 showed no publications by Brazilian authors regarding Antarctica before 1976 on the Scopus database, and only a single publication authored by Aristides Pinto Coelho and published by the National Science Foundation was found on the Web of Science database. The latter was ascribed to the IBEA and was most likely the result of the participation of Pinto Coelho on the US expedition of 1973–1974.

I. J. Cardone, *The Antarctic Politics of Brazil*,
https://doi.org/10.1007/978-3-030-80161-8_3

embassies in the consultative countries to obtain information regarding the institutional design implemented there, their scientific activities and their relations with non-governmental institutions.[2] It also started to have conversations with representatives of the Scientific Committee on Antarctic Research (SCAR) in order to study the possibilities to become an active member.[3] It was precisely as a result of such contacts—and probably under the influence of the example taken from its regional neighbours involved in Antarctica—that the Minister of Foreign Affairs, Antonio Azeredo da Silveira, first suggested the creation of a Brazilian Antarctic Institute.[4]

In order to define a strategy to became engaged in Antarctica, in early July, Minister Silveira suggested the elaboration of a Brazilian policy for Antarctic issues to the Presidency.[5] To ensure efficiency and flexibility, the Minister suggested the task be entrusted to an ad hoc informal working group, which would also be in charge of the design of a Brazilian Antarctic programme and the institutional setting necessary to implement it. The group would be integrated with the Ministry of Foreign Affairs, the National Security Council, the Joint Chiefs of Staff (EMFA), the Ministry of the Navy, the Ministry of the Air, the Ministry of the Army, the National Department of Meteorology, the National Department of Environment, the Ministry of Mines and Energy, the Ministry of Communications, the Secretary of Planning of the Presidency and the

[2] Às Embaixadas em Washington, Buenos Aires, Santiago… (1975). The documentation available reveals the consultations were sent to Brussels, Buenos Aires, Canberra, Moscow, Oslo, Paris, Pretoria, Santiago de Chile, Tokyo, Washington D.C. and Wellington. It is not clear why there is no documented request to the Brazilian Embassy in London. It is probable that both governments were already in consultation on that regard by other channels.

[3] Castro (1975) and Comitê Científico para a Pesquisa Antártica (1975). The Scientific Committee on Antarctic Research is an international scientific organization that coordinate research in Antarctica at the international level. It aroused from the International Geophysical Year of 1957–1958 to study the continuity of the scientific work once the International Geophysical Year was coming to an end.

[4] Silveira (1975a).

[5] Silveira (1975b).

National Research Council (CNPq). Silveira's proposal obtained Presidential approval, and in August of that year, Itamaraty sent letters to those departments[6] requesting them to send representatives.[7]

This proposal was intended to establish the institutional setting, the political principles that would guide Brazil's involvement in Antarctica and the basic lines on which a future Antarctic programme of scientific research would develop. The programme was to be oriented to the training of personnel for polar operations and to develop scientific knowledge in fields with an 'immediate interest for Brazil.'[8] Additionally, it was recommended that there should be cooperation with other active countries in Antarctica after the consideration of political repercussions and the compatibility with 'actual Brazilian interests,' emphasizing the need to prioritize the possibility of collaboration with regional neighbours. Between these fields of interest, as a preliminary basis, the proposal established: (1) the study of maritime currents, particularly those that affect the South American coast; (2) the air currents, especially those that could affect trans-polar flights; (3) the meteorological phenomenon that affect the Brazilian mainland and, particularly, could affect the agriculture and other economic activities; (4) palaeontological studies that could tie the Antarctic continent with South America; (5) the study of Antarctic biota, which could affect the maritime resources of the Brazilian coasts; and (6) the study of the ice-shelves covering the west part of Antarctica and the determination of the causes for its deterioration.[9]

As such, the Brazilian approach to Antarctica followed a pragmatic and utilitarian interest. The scientific involvement in Antarctica was a consequence, above all, of the realization that without an active research programme, participation in the political decisions on the continent would not be possible. Beyond that consideration, the scientific programme to be developed was not actually aimed at studying and

[6] The invitations were directed to: Prof. João Paulo dos Reis Velloso (Planning), Gral. Hugo de Andrade Abreu (Security Council), Dr Alysson Paulinell (Agriculture), Dr Shigeaki Ueki (Mines and Energy), Lt-Brig. Joelmir Campos de Araripe Macedo (Air), Adm. Geraldo Azevedo Henning (Navy), Eng. Maurício Ranjel Reis (Interior), Com. Euclides Quant de Oliveira (Communications), Grl Sylvio Couto Coelho da Frota (Army), Grl Antônio Jorge Corrêa (EMFA).

[7] Grupo de Trabalho sobre Antártida (1975).

[8] Política Antártica Brasileira (1975).

[9] Ibid.

exploring the white continent but should be directed to developing knowledge in those areas which could have a direct impact on Brazil's interests, with an emphasis on the economy. Additionally, international collaboration in Antarctic research was evaluated following political considerations and opportunity, and scientific involvement was relegated to more technical aspects of the implementation of such policy. This seems to have followed the general orientation of the military government installed by the coup of the so-called Revolution of 1964, for which all governmental actions were to be aimed at promoting national interests through a centralized national development programme. Antarctica was seen as a possible means to that end, at least in the potential use of its natural resources.

Between September 1975 and January 1976, the working group elaborated a guideline for the Brazilian National Policy for Antarctica, designated with the acronym POLANTAR, and a draft proposal of such a policy. The guidelines recalled the possibilities opened by the accession to the Treaty and also the additional conditions that should be met in order to participate in the decisions of the regime, in all motivated by:

> Brazil's actual and justified interest in Antarctic affairs, which translates into reasons of national security; given the favourable political, legal, strategic and geographical conditions for an effective Brazilian presence in the Antarctic region; the need for a decisive Brazilian action at the time of the future approval, by consensus, of the definitive status for the region; the possibilities of participating in the exploitation of natural resources of the area and the relevance of knowledge of the physical phenomena originated in Antarctica, especially those with incidence in the terrestrial, maritime and aerial extension of Brazil.[10]

The document revealed the strong influence of the Brazilian geopolitical thought about Antarctica developed in the late 1950s. It repeated some of the arguments advanced some 20 years before by Terezinha de Castro, restating the relevance of Antarctica for Brazil: being the country with the most extensive coast in the South Atlantic Ocean and the inclusion of the continent within the defence zone defined by the Inter-American Treaty of Reciprocal Assistance. Considering the AT as the solely valid

[10] Anteprojeto de política nacional para assuntos antárticos—Subsídios (ca. 1976, p. 2), own translation.

international legal framework for Antarctica and a temporary arrangement,[11] it reiterated that the Treaty guaranteed the non-recognition of any claim, which was compatible with the reservation made opportunely by Brazil[12] in regard to its position 'when the definitive legal territorial status of Antarctica be decided.'

Notably, the document stressed the relevance of the issue of Antarctic mineral resources, considering that the initial stage in which negotiations were at the time allowed Brazil to influence the process. In that regard, the proposal cogitated that the country should try to obtain favourable conditions for the developing countries, as compensation for the difference in technological levels between developed and underdeveloped countries acting in Antarctica.[13]

In terms of the political issues, the document recommended the adoption of a flexible attitude anchored to national interests. Expressing the strong influence that Terezinha de Castro's 'frontage theory' had in their approach to the territorial issue, the guideline suggested the political position:

> ...to be directed towards the establishment of a condominium in Antarctica, in a regime that mainly takes into consideration the geographical and strategic position of those countries which coastlines are predominantly confronted to the Antarctic continent.[14]

That formula was not considered monolithic, and other options, such as the internationalization of the continent, were considered as acceptable. But if the division of the continent was to prevail, Brazil should defend its rights based on its geographical position. Regarding international cooperation, it was instructed to cooperate with the countries already operating in the region, giving priority to the regional neighbours with Antarctic tradition—namely Argentina and Chile—and taking into consideration those investigations that were directly linked to the national interest.

[11] Ibid. In that sense it also repeated the misinterpretation that the Treaty would expire after thirty years after came into force.

[12] It refers to the confidential note sent to the US government in 1958. See Chapter 2 for more details.

[13] Ibid., pp. 3–4.

[14] Ibid., pp. 4–5, own translation.

On the whole, the guidelines recommended the elaboration of a national policy for Antarctica that could ensure a centralized administration of all Brazilian activities of any nature in Antarctica, including public and private bodies; the compatibility of such a policy with the national security and external policies; and the regulation of scientific and natural exploitation activities. It should aim to ensure participation of Brazil in the decision-making process of the Antarctic regime and to establish a physical presence in the region to guarantee a share of any benefit derived from activities on the continent, particularly the use of natural resources. It should also be flexible enough to allow an eventual territorial claim, coordinate all agents involved in Antarctic activities, promote the specialization of personnel on Antarctic issues and promote the development of polar technologies and the use of materials of national origin.

As initial institutional design, the guideline proposed two entities: the *Comissão Nacional para Assuntos Antárticos* (National Commission for Antarctic Issues—CONANTAR) and the *Instituto Antártico Brasileiro* (Brazilian Antarctic Institute—IANTAR). The former was to be attached to the National Security Council and advise the Presidency on the implementation of the POLANTAR. It was to be presided over by the Minister of Foreign Affairs and based in that Ministry, with no additional cost to the Treasury.[15] The IANTAR would be subordinated to the CNPq and would demand only a small expenditure, at least in the initial stage. It would have an executive nature and oversee the implementation of the Antarctic programme, besides being the institution to be associated with the SCAR to represent the country in such scientific forum.

As originally formulated, the guidelines of the POLANTAR envisioned an institutional design that involved a wide range of governmental departments in promoting, executing and administrating Antarctic scientific research in a diverse number of fields. Between the military branches, responsibilities were assigned to several departments: (1) the Navy was to be in charge of the maritime operations, oceanic research and logistic activities; (2) the Army of operating land bases and terrestrial cartography; (3) the Ministry of Air of strategical studies on the aero-spatial importance

[15] It was to be composed by two representatives of the Ministry of the Navy, the Ministry of the Army, the Ministry of Foreign Affairs, the Ministry of Agriculture, the Ministry of Air, the Ministry of Mines and Energy, the Ministry of Planning of the Presidency, the EMFA, the General-Secretary of the National Security Council and the CNPq.

of Antarctica and a military strategy, aerial cartography, meteorological analysis, air support for Antarctic operations and the operation of bases to be assigned; (4) the EMFA of coordinating strategic information on Antarctic activities and research plans of the Armed Forces and the study of a military strategy for Antarctica; and (5) the National Security Council of the study of national security aspects and agreed decisions regarding Antarctica with the Presidency. As well, a number of civil departments were involved: (6) the Ministry of Foreign Affairs was to be in charge of providing administrative and secretary services to the CONANTAR and operate as the political agent in all Antarctic forums; (7) the Ministry of Agriculture of the meteorological and living natural resources studies; (8) the Ministry of Mines and Energy of geological studies and the support for Brazilian mining activities[16]; (9) the Secretary of Planning of the Presidency of equalizing Antarctic activities with the national development plan and provide financial resources; and (10) the CNPq of a wide range of responsibilities related to Antarctic scientific research, including the interchange and transference of technology, specialized assistance in scientific matters, human resources training, scientific events organization and participation and to nominate the director and approve the statutes of the IANTAR.

In addition to the guidelines, a draft proposal of the POLANTAR defined the goals, premises, basic principles, objectives and politics of execution.[17] The document established a centralized character for Brazilian policy in Antarctica, and some emphasis was given to the exploitation of natural resources. Also, it reaffirmed Brazil's interests in Antarctica and their link to national security, its responsibility under the TIAR agreement and the need of active involvement in the region to be included in the political decision-making process. The implementation of the policy was to be in line with the national security policy, the national foreign policy guidelines and should be made compatible with the national development plan. For all that, it was suggested that the Brazilian policy for Antarctica should be flexible and pragmatic and would result in a national Antarctic programme (PROANTAR), which would produce scientific research linked with Brazil's interests and demonstrate

[16] 'Including Petroleum and radioactive minerals.'

[17] Política Nacional para Assuntos Antárticos (ca. 1976a).

an active involvement in the region, guaranteeing an active participation on the regime.

These guidelines and the draft proposal for the POLANTAR were appraised by the National Security Council. The Council's conclusion indicated that the POLANTAR should follow three principles: (1) a sound basis for participation, avoiding any hurried decision-making in implementing a programme; (2) a modest scale; and (3) minimum expenditure.[18] The Council also agreed with Itamaraty's recommendation of resuming the activities of the inter-departmental working group to study the creation of the institutional setting to be implemented—the creation of the CONANTAR and the IANTAR—but considered it more timely to wait until the national policy for Antarctica had been agreed and officially approved.

The different departments involved in the National Security Council appraised the draft version of the POLANTAR, resulting in a slightly modified version that was submitted to the Presidency on 28 October 1976, which was approved the same day.[19] The document remained mostly unaltered with the exception of the incorporation of an opening paragraph, absent in the original draft. The opening echoed the frontage theorists' early formulations by defining the POLANTAR objectives as those of:

> ...the affirmation of Brazilian presence in the South polar continent, in view of possible action, in a near but not immediate future, either to ensure Brazil the benefits of the exploitation of that area, in egalitarian conditions to other nations; either to confront menaces of any nature that, linked to that area, could afflict Brazilian interests, particularly, against the national territory.[20]

[18] Abreu (1976a, p. 4).

[19] Abreu (1976b) and Política Nacional para Assuntos Antárticos (1976b).

[20] Política Nacional para Assuntos Antárticos (1976b), own translation.

Meanwhile, the Brazilian government was evaluating different offers of collaboration,[21] and started to gather information regarding the availability of oil deposits in Antarctica.[22] The opening of discussions about mining activities in the Antarctic Treaty Consultative Meeting (ATCM) in Oslo, in 1975, and the preparatory meeting held in Paris, in 1976, created the sense that the question of natural resources would be soon decided. Nonetheless, the Brazilian government did not prompt any action, remaining interested but keeping its decision of undertaking a policy of gradual involvement.

This was a delicate moment for Brazil, as the conflict about the Itaipú damn was being negotiated with Argentina and the question of Antarctic involvement was being escalated by the press in both countries.[23] As the Itaipú project was more important in the short term, it was natural that Brazil opted not to antagonize Argentina by pressing for a more assertive action.[24] The Argentines had started to conduct their own geological studies of what they claimed as their Antarctic territory, but their technical capabilities for offshore oil exploitation were not developed. The Brazilians were also aware, thanks to their interchanges with the British, that the SCAR had recommended the AT parties to prioritize environmental protection over any commercial interest related to mining activities, and that it questioned whether such activities would be justified from an economic point of view.[25]

[21] The British offered to receive four Brazilian scientists, three in the British Antarctic Survey in Cambridge and one in the Institute of Oceanographic Sciences in London, while the Australians offered to take one Brazilian scientist on their next scientific Antarctic expedition. The government accepted both offers, and two technicians were sent to Great Britain, while a Brazilian officer, the Lieutenant-Commander Luiz Antônio de Carvalho Ferraz (who would play a relevant role in the future development of the Brazilian Antarctic programme), participated on a BAS cruise on the 1976–1977 season. The next year, a further informal invitation was made by the British to send a Brazilian observer for the 1977–1978 season (see: Araujo, 1977; Relações Brasil-Austrália, 1976; Silveira, 1976a, 1976b, 1976c).

[22] Pitão (1976).

[23] See, for example: Antártida Argentina: ¿Qué pretenden los brasileños? (1976) and Teixeira (1976).

[24] Ferreira (2009, p. 235) established a direct link between the delay on undertaking Antarctic activities and the Itaipu damn negotiations, based on the information present in an official document (Aviso 032/1a.SC/122, 18 March 1980, Centro de Documentação do Itamaraty).

[25] Política e Tecnologia (1976).

At the same time, Brazil was following closely the events regarding the announcement by Poland, in 1976, of its intention of establishing an Antarctic base. In October of that year, during the SCAR meeting in Mendoza, Poland's delegation—which was participating as observer—had announced its intention of dispatching a scientific expedition to establish a permanent station on the Antarctic Peninsula. The Polish decision caused some concern in Argentina, Chile and other southern hemisphere countries.[26] The station was to be installed within the territory claimed by those countries and would mean to allow another communist country to be established on the gateway to the American continent. It is important to bear in mind that, at that time, Argentina and Chile were governed by military regimes of an extreme-right and anti-communist ideology. Additionally, the alleged scientific interest of the Polish government clashed with the choice of place for the base, which was in an already crowded region in Antarctica. This raised the suspicions that Poland's interests were mainly commercial and not scientific, and that they were tightly tied to the potential exploitation of natural resources in Antarctica—linked to the harvesting of krill.[27] These suspicions were not without basis, as Poland and West Germany were organizing a southern oceanic research expedition focused on krill, and the Polish even approached Australia suggesting they participate in the expedition.[28]

Poland had wanted membership of the AT even before its signature.[29] It insisted, in early 1958, about its rights to participate of the Antarctic Conference that resulted in the Treaty; it had announced, that same year, its intention to operate an Antarctic station, which it did in the 1959–1960 season; it failed to become a member of the SCAR[30]; and it ratified its accession to the AT even before Argentina, Australia and Chile—that

[26] Following a report from the Brazilian embassy in Buenos Aires, the Polish announcement received a mild acceptance from the northern hemisphere countries, for which the Antarctic politics should be subordinated to the scientific interests. In contrast, in Argentina, Australia, Chile and South Africa, the announcement caused some astonishment. See: Souza (1976b).

[27] Machowski (1991, p. 68).

[28] See: Krill and Other Marine Activities (1976).

[29] For early Poland's intends to access to the Antarctic Treaty and its further involvement see: Machowski (1991).

[30] The SCAR decided, on its third meeting in Canberra, 1959, that Poland could be accepted '...from the date of the disembarkment of their wintering party' (SCAR, 1959, p. 591).

is, even before the Treaty came into force. But it was the prospects of commercial exploitation what seems to have drove Poland to resume their efforts in Antarctica.

In any case, for Brazil, Poland's action presented an opportunity. If Poland was to establish itself in the sector claimed by Argentina and Chile, it would make easier for Brazil to install its own station in that region. It was expected that Argentina and Chile would see it to be more agreeable to have a regional partner than a communist country and thus provided their explicit support to the Brazilian initiative.[31]

Poland's expedition and the installation of the Professor Henryk Arctowski station on King George Island (*25 de mayo/Vaterloo*) forced the AT consultative members to consider the issue of accession to consultative status. In the first Special ATCM, held in London in 1977, the twelve original members decided in favour of Poland's request to become a consultative party. The decision was based on the previous season's Polish expedition and the installation of a permanent base. Nonetheless, misgivings from some of the parties did not dissipate.

In terms of the regime, the need to find some political solution to the Polish decision to set foot in Antarctica forced the parties to reach a more explicit formula of accession than the general provision of conduct 'substantial' scientific research activities. After deliberations, the formula included successive steps: (1) notification to the US government as depositary of the Treaty of the intention to access to consultative status along with information regarding the Antarctic activities in which such request was based; (2) evaluation by the AT consultative parties of the merits of the request and the adoption by the candidate of the recommendations already agreed by the parties; (3) convene a Special Consultative Meeting; (4) adoption of the decision in the Special Consultative Meeting, which could only grant consultative status by unanimity.[32]

While the status of Poland was about to be discussed, the interministerial working group on Antarctic issues was submitting their second report to the Brazilian Presidency. The group had resumed their activities in late January 1977, after the POLANTAR was approved in late October of the previous year, and extended their work until the end of

[31] See, for example: Silveira (1977a) and Souza (1976a).

[32] Final Report of the First Special Antarctic Treaty Consultative Meeting (1977).

March. Their updated objectives were to establish the specific competences for the different governmental departments involved in future Antarctic activities, including the elaboration of projects for the creation of new institutions specifically focused on Antarctic activities and the draft of a press release. The latter was aimed to avoid misinterpretations and speculations about the intentions of the Brazilian government.[33]

The new working group included representatives from similar ministries.[34] In terms of functions and responsibilities, the new document did not alter in any relevant way the previous one and reflected the same interest in the potential use of natural resources. As well, the draft version of decree to create the CONANTAR and the IANTAR incorporated most of the previously agreed functions and competences.[35]

By its part, the press release restated the:

> ...firm decision of the Brazilian government of maintaining an active participation in Antarctica, as resultant of the character of Brazil as a country directly and substantially interested in Antarctica...[36]

The release also affirmed the interest demonstrated by different sectors of Brazilian society, such as debates of the National Congress, the press, the Historical and Geographical Institute, the EMFA, the National Security Council, the War College, and the analyses of specialists of national private institutions. Additionally, it reported the decision of creating the CONANTAR and the IANTAR, articulating the Antarctic activity at the domestic level to the national interests, and at the international level with the rights and obligations established by the AT.

With some consensus about the institutional design and with the perspectives open to accession by the incorporation of Poland as consultative member, the Brazilian government looked for support to be accepted

[33] Silveira (1977b).

[34] The only exceptions were the incorporation of the Ministry of Economy and the exclusion of the Ministry of Interior and the Ministry of Communications.

[35] The only exception is the incorporation to the CONANTAR the function of 'coordinating the national presence in meetings, congresses and working groups or any other activity linked with technical and scientific issues related with Antarctica,' which was previously in charge of the IANTAR (Art. 4—paragraph g of the Draft Decree. Enclosure to Silveira, 1977b).

[36] Projeto de Comunicado à Imprensa. Paragraph 2. Enclosure to Silveira (1977b), own translation.

as observer in the 1977 ATCM. The efforts taken before the UK government obtained a favourable predisposition, leading the Brazilian government to ask for support from Argentina and Chile. The two Latin-American countries expressed their support for the Brazilian initiative, looking with approval at the information, provided informally, that the Brazilian government wanted to create a scientific institution to deal with its Antarctic programme.[37] However, the effort did not come to fruition, probably because of the fear, between the consultative parties, that such an avenue could open the possibility of a wider intervention in Antarctic affairs.[38]

As the documentation available reveals, Itamaraty was following with interest the developments related to Antarctica and its natural resources. The potential exploitation of oil was added to the manifested interest expressed by the scientific expedition of Poland and West Germany aimed to study krill and other alleged interests in natural resources by the US, Russia and Japan.[39]

While Brazil's participation in the ATCM did not materialize, it was authorized to participate as observer in the XV Meeting of the SCAR in Chamonix.[40] The Brazilian representative was joined by representatives from Poland, the Federal Republic of Germany, the German Democratic Republic and the Intergovernmental Oceanographic Commission as observers at the meeting.[41] On this occasion, Poland and the Federal Republic of Germany were accepted as members of the SCAR and discussions about the potential effects of exploitation of natural resources, living and non-living, continued. For Brazil, participation as observer on the SCAR meeting was far from satisfying its desire of to be included in the discussions about the future use of natural resources in Antarctica. However, it was the first step to being informed about the negotiations that were taking place.

[37] IX Reunião Consultiva (1977).

[38] Ferreira (2009, p. 136).

[39] See the correspondence dated in 1977 and 1978 present in: Antártida. – Dossiê (BR DFANBSB Z4.REX.IPS.27), Arquivo Nacional, Brazil.

[40] The Brazilian representative was Ambassador Cannabrava, who was secretary of the inter-ministerial working group.

[41] Scientific Committee on Antarctc Research (SCAR) (1978).

By that time, discussion about the implementation of the Biological Investigations of Marine Antarctic Systems and Stocks programme (BIOMASS),[42] in charge of providing information about the Southern Ocean ecosystem to ensure its proper management and administration, was well advanced. The 1977 London ATCM had recommended that the parties should collaborate in scientific research on marine living resources, and in 1978, the Conference on the Conservation of Antarctic Marine Living Resources was established.

Reports on the potential environmental impacts of exploitation of mineral resources were also being produced. The issue of natural resources was becoming an important challenge to the Antarctic Treaty System (ATS), and the SCAR reacted by warning about the need of protecting the Antarctic environment, a position the southern countries shared.

Negotiations on marine living resources adopted a character of urgency due to Poland, USSR and Japan's declared interests in fishing for krill, and overtook the discussion about mineral resources.[43] The debates extended for more than two years in a process in which participation was limited to AT consultative members, no observer was accepted, and the content of the formal and informal meetings was kept in secret.[44]

Despite the closed nature of the negotiations and the alleged interest of Brazil in the futures of the continent, there is no evidence that its government had protested or tried, in any way, to influence or alter the outcome of such negotiations. Unlike the interest shown by Brazil in being involved in any negotiation regarding Antarctic mining activities, the issue of marine living resources did not seem to have motivated any

[42] Discussions about the potential effects of krill exploitation started in 1972, when a group of experts was established by the SCAR. That group was cosponsored by the Scientific Committee on Oceanic Research (SCOR), the International Association for Biological Oceanography (IABO) and the Advisory Committee on Marine Resources Research (ACMRR) of the Food and Agriculture Organization (FAO). In 1976, the BIOMASS programme was established in order to inform the decision-making process. For more details on the BIOMASS programme and its effects on the ATS, see: El-Sayed (1991).

[43] Beck (1986, p. 178).

[44] Ibid., pp. 178–179.

reaction from its government. This is probably because of a lack of defined benefits, at that time, of fishing in southern waters could bring.[45]

In the meantime, the IBEA continued to exert its influence and pushed for an Antarctic expedition of its own. In April 1977, the IBEA participated in an international congress in Punta Arenas, where the issue of development of Antarctic resources was discussed.[46] Along with prominent Antarctic researchers from Argentina, Australia, Chile, Japan, New Zealand, Norway, United Kingdom and the United States, Brazilian participation was increased by official delegates from the Ministry of Foreign Affairs, the CNPq and the Navy.[47] The participation of the IBEA in such a forum was an important international acknowledgement that could be seized for supporting the institute's upcoming expedition. However, whether these developments seemed to give support to the IBEA as institution, the lack of official backing and internal divisions conspired against the advancement of the project.

In 1979, with the change to the Presidency in Brazil, the steering of its foreign policy experienced an important change, in particular with Argentina. This would have important repercussions for its Antarctic involvement. Figueiredo's assumption to the Presidency was characterized by a rapprochement with Argentina, reaching an agreement over the long-lasting Itaipu dam dispute, and thus solving the most important conflict between the two neighbouring countries.[48]

The dispute, which had marked the relationship between the two countries for much of the 1970s, had begun to show signs of distention in

[45] During the research it was only found a single article published in 1981 in the magazine '*Manchete*' (Krill/O Alimento do Futuro, 1981) that covered the potentialities of the krill for the alimentation industry. All other mentions to the resource are merely incidental.

[46] '*Symposium internacional sobre el Desarrollo de la Antártica*,' organized by the International Studies Institute of the University of Chile and the Institute of the Patagonia. Participation on such event was reflected on the volume '*El desarrollo de la Antártica*' (Vicuña & Araya, 1977), with contributions by Wiltgen and Pinto Coelho from the IBEA.

[47] The delegation included Ambassador Cannabrava (Itamaraty); Dr Dourimar Nunes de Souza (CNPq); and Lieutenant-commander Luiz Antônio de Carvalho Ferraz (Brazilian Navy), who participated as observer in the British Antarctic Expedition of 1976 on board of the *Endurance* and the *Bransfield*.

[48] Spektor (2002).

1977, reaching, in October of the following year, a preliminary understanding that made it appear that an agreement was near.[49] However, new misunderstandings between the Argentine and Brazilian delegations foiled the agreement and increased the Brazilian annoyance—especially that of its Chancellor, Silveira—up to the end of the Geisel's government. It was only after Jõao Batista Figueiredo, a general in the Brazilian Army who served during Geisel's Presidency as Head of the National Intelligence Service (*Serviço Nacional de Inteligencia*),[50] assumed the Presidency that a new attitude for Argentina was adopted and the possibility of an Agreement became possible.

Figueriedo personal visits to Buenos Aires demonstrated a commitment to establishing a closer relationship between both countries; while, under his command, the Brazilian diplomacy was opening the way for cooperation in the military, nuclear and economic fields.[51] This rapprochement with Buenos Aires was built on new principles, different from the traditional 'official cordiality' (*cordialidade oficial*) that characterized the previous periods of close relationship. Brazil did not want to antagonize Argentina, which it saw as a natural regional partner. However, due to its heightened status in Latin-America and the international arena, Brazil gained the confidence to defend its position despite possible opposition from Buenos Aires.

That redefined situation with Argentina overthrew the last barrier—the Itaipu dam dispute—that in the past had motivated reluctance within the government to launch an Antarctic expedition. Additionally, this coincided with the increased urgency that the issue of natural resources was acquiring within the ATS and with advancements in the organization of the Brazilian Antarctic institutional setting. In July 1980, the EMFA wrote to the Presidency to draw attention to the need of reactivating the inter-ministerial working group in order to foster the fulfilment of the POLANTAR objectives, taking advantage of the good relations with Argentina.[52] The document stated the need to abandon the outdated

[49] Ibid., p. 134.

[50] It was also who required a report on Antarctica to the EMFA in 1971, when being head of the Military Cabinet. See: Section 2.2.

[51] Spektor (2002, p. 136).

[52] Ferreira (2009, p. 132).

territorial thesis based on the frontage theory and to obtain a consultative status to defend its interest within the regime.

Outside governmental circles, the IBEA continued to pursue its aim to organize an Antarctic scientific expedition and looked for support from the neighbouring country of Uruguay. In January 1981, the IBEA President Wiltgen travelled to Chui, on the Brazilian border with Uruguay, to meet Pericles Azambuja, a journalist who had favoured the cause of Brazil's involvement in Antarctica for many years and with whom he had established a recent partnership to promote the IBEA expedition.[53] From there, they travelled to Montevideo and Las Flores to meet Uruguayan personalities involved in promoting an Antarctic expedition. Julio Musso, founder of the Uruguayan Antarctic Institute (*Instituto Antártico Uruguayo, IAU*), was amongst that group. The intention of the meeting was to study the possibilities of organizing a joint Brazilian-Uruguayan expedition, which the Uruguayans were keen on. However, the project aroused some misgivings in Musso, who believed that it wouldn't be seen with enthusiasm by the military government–ruling Uruguay at that time and which had removed him and all other civilians from the IAU.

Thanks to the conversations, the group planned a joint expedition, which would follow the principles established by the SCAR and would go to the Norvegia Cape.[54] The plan was divided into four phases from the recognition of the region, to the establishment of a fixed station, ports and an airfield. That would be complemented with other stations, including one on Bouvet Island, to establish a synoptic network of observations.

That first contact was followed by fertile discussions between the two groups, coupled with a press campaign to exert pressure on both governments. The Uruguayan government misgivings about Musso, who

[53] This and the following are based on Azambuja's accounts of the events. See: Azambuja (2005, Chapter 4). Whether Azambuja's book reflects a highly-coloured account of the events and, particularly, his own participation, it also provides some interesting details about some undocumented events and, thus, represents a valuable source of information.

[54] Azambuja states that the choice of place was linked with the position of Brazil and Uruguay within the 'theory of sectors' (Azambuja, 2005, p. 45). However, the Cape Norvegia (actually, Azambuja refers to it as Cabo Norvega) is located outside the whole 'American Sector' and, thus, outside the claimed sectors defended by the frontage theory. Furthermore, Musso declared that such joint enterprise would be against any type of territorial claim (Fontes, 2012, p. 5).

was considered an anarchist, and the political inconvenience of joining a project that could antagonize the Argentines made them downplay Musso's role and ended the idea of a joint venture.[55] The corroboration that the IBEA was not an official entity deprived the enterprise from any attractiveness for the Uruguayan government, while there was a high risk of producing a strong reaction in Argentina, as was already happening through the press.

Whether the Argentine government had expressed some mild form of support for Brazil's participation in the Antarctic regime, some officials had expressed concerns on the new form of Brazilian 'imperialism' in Antarctica. Most notably, Jorge Leal, an Argentine Army general—who rose to fame in Argentina for being the first national to reach the South Pole by land in 1962—expressed concerns that Brazil was colluding with the US Government to advance on the oil exploitation of Antarctica.[56]

Notwithstanding the misgivings in some sectors of Argentina, in August 1981, the National Security Council recommended resuming the work of the inter-ministerial working group to amend the POLANTAR with respect to competences and skills; create the CONANTAR and the IANTAR; and promote the participation of Brazilian 'technicians' in Argentina's and Chile's expeditions to the continent in the upcoming seasons.[57]

Taking advantage of the impulse that the Antarctic cause was gaining in Rio Grande do Sul, Wiltgen and Azambuja promoted the creation of a branch in that State, which was formalized on 12 September 1981. The location of Rio Grande do Sul, the southernmost state of Brazil, held a symbolic appeal for leading the Antarctic enterprise. Additionally, it was anticipated that if some form of exploitation of the Antarctic were to happen—as it was the expectation in Brazil at that time—the southern state would benefit from effective Brazilian involvement.

On 10 October 1981, the IBEA received the 'declaration of public utility' and, at the end of the month, was informed by the Ministry of the

[55] Fontes (2012, pp. 20–21).

[56] Azambuja (2005, p. 62) and Fontes (2012, p. 22).

[57] Aspectos da Conjuntura Política Nacional – Análise N°3 (1982, p. 30).

Navy about the Government's intention to dispatch an Antarctic expedition.[58] The IBEA branch in Rio Grande do Sul rapidly started to prepare a proposal to be submitted to the Navy, which included participation from the *Universidade Federal do Rio Grande* (FURG) and the *Universidade do Vale do Rio dos Sinos* (UNISINOS). Nonetheless, they did not wait until the Brazilian official expedition. After obtaining the green light from the Chilean government, a group of five scientists departed as guests, in February 1982, to Antarctica.[59] During their stay on the continent, which lasted until the end of March, scientific observations were made and they obtained first-hand experience with the materials and equipment.[60]

Lacking any official character, participation as guests of the Chilean operations in the region did not differ significantly from the previous experiences, but it was an advance of what came the following months. The decision of the government to go to Antarctica with an official expedition was taken and preparations were made. Finally, after 10 years of efforts, the Antarctic dream of the IBEA was coming true. Later, once Brazil reached Antarctica and acknowledging the relevance that the IBEA had in that achievement, the Brazilian Ministry of the Navy, Admiral Fonseca, said: 'Thanks to the IBEA, Brazil is in Antarctica and, for that reason, the Navy's doors will be always open to the IBEA.'[61]

3.2 And Thus, the Tropic Met the Pole

Since Brazil's accession to the AT in 1975, participation by Brazilian naval officials in foreign expeditions to Antarctica served as a first step in acquiring the necessary knowledge and experience to organize an official campaign operated by the Navy. One of those officials, Luiz Antônio

[58] The information in this paragraph has been obtained from Azambuja (2005, pp. 30; 61–86). Azambuja interprets that the decision of the Government and the Navy was forced by the events in Rio Grande do Sul. Despite the IBEA project had exerted an important influence through lobby and the public opinion, during the research no evidence was found to support the idea that the final decision to organize the expedition had any direct relation with the events in Rio Grande do Sul, and there are good reasons to believe that the decision was adopted in advance and was only communicated more broadly at that opportunity. See, for example: Bakker (2004).

[59] It is not clear if that possibility was negotiated by the Brazilian Navy after the recommendation made in August by the National Security Council.

[60] Azambuja (2005, pp. 80–82).

[61] Fonseca, cited by Pinto Coelho (1984, p. 81, own translation).

Ferraz, who was commissioned to represent Brazil in several international scientific forums related to oceanography and hydrography—including the seminar that took place in Punta Arenas, Chile, in 1977—was essential in promoting the organization of Brazil's first Antarctic expedition. Ferraz was instrumental not only due to his knowledge of oceanographic science and his belief in the importance of the scientific endeavour, but also due to his international links, including his relationship with British officials thanks to his participation in the 1976 British Antarctic Survey's expedition on board of the *Endurance* and the *Bransfield*.[62]

Another key figure in promoting Brazil's Antarctic involvement was Admiral Maximiano Eduardo da Silva Fonseca. Fonseca was a rising officer within the Navy, with extensive experience in oceanographic work, when he was appointed Minister of the Navy by the President Figueiredo, in 1979. As Minister, Fonseca promoted the incorporation of naval capabilities for the Brazilian Navy. He was responsible for the creation of the Brazilian Navy nuclear programme, the acquisition and construction of submarines and the extension of the scientific work as a way to consolidate Brazil's maritime rights.[63] Since the 1970s, Brazil had embarked on a campaign to extend its territorial sea to 200 miles—as had already been done by Argentina and Uruguay—which was seen as essential to protect its economic interests. Whether Brazil had, up until then, projected as a continental state, the realization of the potential of oil exploration on the maritime platform drove some leaders, and particularly the Navy, to promote the idea of a 'blue Amazon'—that is, as a region of natural wealth, parallel to the green Amazonia in extension and richness, that belonged to Brazil by right of sovereignty.[64] However, Brazil lacked the resources to control such an extended region, and Admiral Fonseca was well aware of those limitations.

[62] Informations obtained in the Brazilian Navy official website. Source: https://proantar.com.br/page/capitao-de-fragata-luiz-antonio-de-carvalho-ferraz-1567689830. Accessed on: 8 November 2020.

[63] The biographic information of Adm. Maximiano Fonseca was obtained from the Brazilian Navy oficial website. Source: https://www.marinha.mil.br/dphdm/sites/www.marinha.mil.br.dphdm/files/almiranteDeEsquadraMaximianoEduardodaSilvaFonseca.pdf. Accessed on: 10 November 2020.

[64] Marinho (2016).

However, this situation was coupled with an economic crisis that made necessary a difficult balance between financial resources and the implementation of a policy that could offer some long-term solutions to Brazil's structural deficits. The exhaustion of the economic and productive capacities of the first stage of the military dictatorship that produced the so-called economic miracle led, from the late 1970s, to economic stagnation, rising inflation and a foreign debt crisis, parallel to what was happening in almost all Latin-American countries.[65] One possible solution was the development of natural resources that could provide the necessary boost to the economy by reducing the trade and fiscal deficit. In this regard, the sea-bed and its natural resources, especially hydrocarbons, offered great prospects. This was framed by the discussions of the Third United Nations Convention on the Law of the Sea (UNCLOS) that had been taking place since 1973 and in which Brazil defended the position of the extended territorial sea of 200 miles.

With the intention to promote a more dynamic response from the Brazilian government to the issue of the territorial sea, Adm. Fonseca re-structured the *Comissão Interministerial para or Recursos do Mar* (Inter-Ministerial Commission for Maritime Resources—CIRM). The commission was created in 1974 to promote the safeguard of Brazil's economic interest in the maritime platform but, in the view of the Minister, lacked dynamism. To favour continuous work, Fonseca created, in 1979, the CIRM Secretariat, which would have a pivotal role in the promotion and organization of the first official Brazilian Antarctic expedition.

In June 1981, Rear-Admiral Múcio Piragibe Ribeiro de Bakker was assigned the post of Secretary of the CIRM and immediately started to campaign for an Antarctic expedition under the auspices of the CIRM, taking advantage of his contacts with other relevant authorities within the Ministry and the EMFA.[66] With the approval of the upper hierarchy of the Navy, Bakker started to push forward the plans for dispatching an official expedition as soon as possible. Bakker met

[65] The reference is to the debt crisis of the 1980s known as the 'lost decade.'

[66] This and the following account are based on Bakker (2004, pp. 85–87). By the accounts of Bakker and analysing the professional trajectory of almost all relevant names of the Navy that were, in one way or the other, involved in the first OPERANTAR, it seems that the Diretoria de Hidrografia e Navegação (Directorate of Hydrography and Navigation) was key in generating interest in Antarctica within the Naval ranks.

Ambassador Rubens Ricúpero, by then Chief of the Department of the Americas, and Professor Lynaldo Cavalcanti de Albuquerque, president of the CNPq, to propose that the CIRM would be in charge of coordinating an Antarctic expedition. The proposal generated some reluctance, being seen as though the Navy was taking control of an expedition the main objective of which was scientific. However, thanks to the fact that the CIRM included representatives of other ministries and that it would facilitate the dispatch of the expedition in a short timeframe, it was finally accepted.

With no polar experience or equipment, the decision of dispatching an official Antarctic expedition under the Brazilian Navy required the acquisition of a vessel capable of navigating the ice conditions. Negotiations started with the British for the acquisition of the *Endurance*, an ice-patrol vessel that was on the list of ships to be decommissioned in 1982 as part of the planned cuts in defence spending by the government of Margaret Thatcher.[67] The *Endurance* was known not only to Ferraz, who participated in its 1976 campaign, but also by Minister Fonseca.

By October 1981, negotiations for the *Endurance* were well advanced, but opposition within the British Parliament stalled the selling of the ship.[68] That led Brazil to study other options, like the acquisition of a polar ship from Poland or the transference of a US decommissioned polar vessel *Eltanin*, which served between 1974 and 1979 on the Argentine Antarctic fleet as *Islas Orcadas*.[69] The Brazilian government was determined to dispatch an expedition to the south even if it had to limit its research in the sub-Antarctic region with the existing oceanographic ships. However, the acquisition of a polar vessel in order to reach the Antarctic

[67] The policy was part of the Defence White Paper presented by Thatcher's government in June 1981 under the leadership of Secretary of State for Defence, John Nott: 'The UK Defence Programme: The Way Forward' Cmnd 8288 (1981). The ship would eventually have a relevant role during the next year Malvinas/Falklads war, operating in the South Georgia.

[68] Berg (1981).

[69] Polônia reduz expedições (1981), Brasil prepara missão (1982), and Buque hidrográfico A.R.A. (n.d.). *Eltanin* was the same vessel the IBEA manifested interest in 1972–1973 (see Section 2.2).

region was considered essential to acquire consultative status in the AT, which was the main aim of its Antarctic policy.[70]

By the end of 1981, Itamaraty insisted on the need to promptly dispatch an Antarctic expedition.[71] Following the report delivered in August, in December, the National Security Council also insisted on the need to dispatch a scientific expedition to Antarctica coordinated by the Navy.[72] In order to maximize the efficiency on the use of resources, the idea of creating an Antarctic Institute was postponed and its responsibilities attributed to the CIRM, as suggested by Bakker. To elaborate and coordinate a programme for the expedition, a subcommittee of the PROANTAR within the CIRM Secretariat was created, under the direction of Lieutenant Commander Eugênio Ferreira Neiva.[73]

Despite the decision of assigning the responsibility to the CIRM being based on financial concerns, the choice was an expression of how Brazil saw Antarctica: as another potential source of natural resources linked to maritime exploitation. As was in the case of the expansion of the Navy and the research linked with the potential use of natural resources from the maritime platform,[74] research and the acquisition of polar capacities by the Navy were linked with the future prospects of natural resources exploitation—especially oil—and thus the CIRM appeared as the natural choice.[75] For the Navy, the operation in Antarctica would also result in the acquisition of capabilities for overseas projection of national power

[70] This was specially the case when its participation on the 1958 preparatory negotiations for the Antarctic Treaty was rejected under the grounds that it has not conducted research on the Antarctic region.

[71] Ferreira (2009, pp. 132–133).

[72] Aspectos da Conjuntura Política Nacional - Análise N°3 (1982, p. 30).

[73] Bakker (2004, p. 87).

[74] Marinho (2016).

[75] The Brazilian representation at the UN General Assembly declared in 1984 that '...the choice of the CIRM did not imply that the country's Antarctic programme would be oriented mainly towards research on resources of the sea' (Question of Antarctica, 1984, pp. 4–5). Whether that could be true, at least to some extent, as regards the research in Antarctica, the way in which Brazil defined its interest in Antarctica, the formulation and objectives included in the POLANTAR and the discourse presented continuously in Brazil—even to justify the expenditure in scientific research—suggest that the prospects of natural resources have been one of the key drivers of its involvement in Antarctica.

in a region that was considered as having strategic importance.[76] In all, that resulted on the Navy having a pivotal role in Brazil's involvement in Antarctica,[77] something that would structure the shape and operation of its Antarctic programme.

Following that decision, on 12 January 1982, two decrees were issued: N° 86.829/82 established the *Comissão Nacional para Assuntos Antárticos* (National Committee for Antarctic Issues—*CONANTAR*), and N° 86.830/82 assigned the responsibility for the elaboration and execution of the Antarctic Programme to the CIRM, pending the creation of 'specific institutions for the execution of the Brazilian National Policy for Antarctica'[78]; the latter would not eventuate. The CONANTAR mainly inherited the composition of the inter-ministerial working group[79] and was assigned with the main political responsibility, advising the President and being in charge of defining the guidelines for the National Policy for Antarctica; evaluating and approving the National Antarctic Programme; coordinating all issues related to Antarctica; suggesting the allocation of financial resources for the development of Antarctic activities; promoting the involvement of universities and research centres in Antarctic research; promoting the updating of legislation related to Antarctica; and coordinating the participation of Brazilian nationals in scientific events. The CIRM would be subordinated to the CONANTAR in the implementation of the Antarctic programme and would be subject to the guidelines defined by such committee.

[76] This was recognized by Adm. Fonseca in a discourse given in April 1982, when he stated that: 'In what relates to the Navy, it should be noted that its effective involvement in the Brazilian Antarctic programme (PROANTAR) will stimulate its oceanic features, letting it to obtain knowledge and operational capabilities in an area of fundamental strategic importance' (Fonseca, as cited in L. F. de Mattos, 2015, own translation).

[77] Despite the alleged inter-ministerial character of the CIRM, the centrality of the Navy was completely apparent—and it still is. See: Fontes Diégues (1987, p. 100).

[78] Decreto N° 86.830 (1982), own translation.

[79] It included the Ministry of the Navy, the Minister of the Army, the Ministry of Foreign Affairs, the Ministry of Agriculture, the Ministry of the Air, the Ministry of Mines and Energy, the Secretary of Planning of the Presidency, the General Secretary of the National Security Council, the EMFA and the CNPq. Just a few months later, in May 1982, the Decree 87.217/82 added the Ministry of Education and Culture, most probably as a way to study the possibility of the inclution of content related with Antarctica in the educational curriculum, something that Argentina and Chile were since long doing as part of their national policies for Antarctica.

With the institutional setting in place, everything was ready to advance on the concretization of Brazil's presence on the white continent. However, other developments were taking place in the South Atlantic. The Malvinas/Falklands War between Argentina and the United Kingdom over the South Atlantic archipelago broke out on 2 April 1982 and added new obstacle to the Antarctic plans of Brazil. Whether the Brazilian government had any answer or not before the rise of tensions between the Argentines and the British over South Georgia and Malvinas/Falklands at the end of March 1982, the *Endurance* had a relevant role during the war and thus it is unlikely that the selling the ship would have advanced any further. Moreover, the theatre of conflict was in the middle of the route from Brazil to Antarctica, which could hamper—or at least imply an important detour—the expedition's way south. The rapid conclusion of the war, with the capitulation of the Argentines by the middle of June, allowed the Brazilian government to continue with its plans.[80]

Notwithstanding the conflict on the Malvinas, in May 1982 a seminar on Antarctic issues took place on the Oceanographic Institute of the University of São Paulo. The event provided the opportunity for scientists, members of the CIRM subcommittee and foreign guests to exchange information and to outline the project of the Brazilian Antarctic programme in a more concrete way.[81]

Despite the failure of the negotiations with the British, by July 1982 Brazil had secured the purchase, for USD 3.05 million, of the *Thala Dan*, a Danish polar vessel that had previously been freighted by the Australian Antarctic programme.[82] With the polar vessel secured, the plans for conducting sub-Antarctic activities also advanced. By then, it was expected that the *Thala Dan* would be joined by the oceanographic vessels *Almirante Câmara,* of the Brazilian Navy, and *Prof. W. Besnard*,

[80] Ferreira states that the war delayed the meeting of the first CONANTAR (Ferreira, 2009, p. 133). However, the research revealed the Commission met informally three times during 1982 (see: Aspectos da Conjuntura Política Nacional – Análise N°5, 1982, p. 70).

[81] Bakker (2004, p. 88).

[82] The *Thala Dan* had a length of 82.11 m, a beam of 13.72 m, displacing 2.182 tons and reaching speeds of 12.5kt. It had the capacity for seventy-seven people (crew included) and space for two helicopters.

of the Oceanographic Institute of the University of São Paulo,[83] but only the latter was included in the final programme for the PROANTAR.

The plans were advancing at a good pace and the possibilities of launching the expedition for the 1982–1983 season seemed challenging, but possible. On 11 August 1982, the subcommittee for the PROANTAR lost one of its key members, Commander Ferraz died of a heart attack while representing the Brazilian Navy at the 3rd Joint Oceanographic Assembly in Halifax, Canada. The regretful loss of Ferraz was a heavy blown for the Brazilian Navy but it did not change the course of events. Since August, and taking advantage of the contributions made in the seminar, the CIRM drafted a scientific programme organized in five sub-programmes that could be incorporated in the activities to be developed during the First National Antarctic Expedition (OPERANTAR I): Life Sciences; Geosciences; Atmospheric Sciences; Logistics; and Education and Training. In October, the programme was approved by the CIRM.[84]

The polar vessel *Thala Dan*, re-christened *Barão de Teffé*[85] by the Brazilian Navy, was delivered in September 1982, just three months before the expedition.[86] As part of the purchasing contract, two Navy officials were sent in advance to join a voyage to Greenland to train to navigate in icy waters.[87] Also in September, the chief scientists of the upcoming expedition joined the third meeting of the BIOMASS technical group in Germany and arranged the collaboration with the Second International BIOMASS experiment, which would be eventually conducted on the *Prof. Besnard* in the Bransfield Strait (*Mar de la Flota*), between the South Shetlands and the Antarctic Peninsula.

On 7 October 1982, the CONANTAR had it first official meeting. In the opening address, the Minister of Foreign Affairs, Ramiro Elysio Saraiva Guerreiro, commented on the importance that Antarctica had for Brazil in terms of security and economic concerns. The Minister also

[83] Brasil prepara expedições (1982) and Aspectos da Conjuntura Política Nacional – Análise N°5, p. 70. The *Prof. Besnard* displaced 703 tons, reaching speeds of 12kts, had the capacity for twenty-two people (*Brasil na Antártica: 25 anos de história*, 2008, p. 15).

[84] Aspectos da Conjuntura Política Nacional – Análise N°5 and Bakker (2004, p. 88).

[85] The ship was named after Admiral Antônio Luís von Hoonholtz (1837–1931), Barão de Teffé. Hoonholtz was a Brazilian diplomat, geographer and politician who was commissioned with the study of Venus during the First International Polar Year.

[86] *Brasil na Antártica: 25 anos de história* (2008, p. 15).

[87] Henriques (1984, p. 111).

stressed the cooperation offered by Argentina and Chile and asserted that cooperation with Peru and Uruguay, who were also organizing their own Antarctic programmes, would be pursued. In addition, making emphasis on the natural resources and environmental links, the Minister highlighted that:

> Brazil has direct and substantial interests in the Antarctic, being evident through a simple glance at the map. The Southern Continent is the origin of multiple phenomena, especially meteorological ones, that directly and strongly influence our climate, with relevant economic effects. The fishery resources that proliferate in the waters adjacent to the sixth continent are already recognized as a source of proteins that could play a significant role in solving the food problem globally. The progress of the earth sciences today allows us to assess the geological relation between Antarctica and South America and Africa, and to infer, even when not knowing exactly, the presence of deposits of solid minerals and hydrocarbon resources.[88]

On 18 December 1982, President Figueiredo, accompanied by Minister Fonseca, Secretary Bakker, Minister Saraiva Guerreiro and other authorities, visited the *Teffé* in the port of Rio de Janeiro, officially giving the farewell to the first official Brazilian Antarctic Expedition.[89] The *Teffé* set sail on the morning of the 20 December 1982, a century after the Brazilian expedition that joined the study of the transit of Venus in Punta Arenas.[90] On board were 86 people, including the crew, members of the naval aviation, reporters, scientists,[91] a representative from the Ministry of Foreign Affairs and four foreign observers: one Argentine, two Chilean and one Peruvian naval officials.[92] It also counted on the assistance of a Danish pilot with experience in Antarctic navigation and with two alpinists from the *Clube Alpino Paulista* (São Paulo Alpinist Club) who would

[88] Saraiva Guerreiro (1982, p. 73), own translation.

[89] Nunes (2005, p. 111).

[90] The reference here is to the Imperial expedition sent to study the transit of Venus on board the vessel *Parnaiba* under the command of Admiral Hoonholtz, the Barão de Teffé.

[91] The only five women in the expedition were in this group.

[92] The Foreign Ministry representative was Conselheiro Luiz Filipe de Macedo Soares, who would eventually be the first director of the Sea, Antarctic and Space Division of the Ministry of Foreign Affairs (L. F. de Mattos, 2015, n. 125). A complete list is present in: Nunes (2005).

assist with the land operations,[93,94] establishing a relationship that would endure up to today. To support on the navigation and logistics of the expedition, the *Teffé* embarked two Westland Wasp helicopters.

The *Teffé* was joined by the oceanographic vessel *Professor Besnard*, which departed the same day from the port of Santos, in São Paulo.[95] Both vessels called at the port of Rio Grande, where they stayed for three days to complete preparations before heading south.[96] The main objective of the *Teffé* was prospecting potential settlements for a Brazilian station. Its navigation plan included a visit to the tip of the Antarctic Peninsula and the Gerlache Strait, a call in at Punta Arenas for refitting, and the visit of the West German Neumayer station, east of the Weddell Sea, before heading back to Brazil. For its part, the *Besnard* would mainly be dedicated to scientific activities focused on the study of krill, in order to train in the required techniques and competences to collaborate with the BIOMASS programme, to which it had committed a few months earlier in Germany.[97]

The Brazilians were aware of the challenge implied in crossing the Drake Passage[98] but they did not expect to encounter any other critical events on their way south. However, despite the war between the Argentines and the British having ended some six months earlier, the Brazilian expedition suffered the effects of the tensions that remained. On its way to Antarctica, when crossing international waters at the latitude of the Malvinas/Falkland Islands, the *Teffé* was harassed by British forces. On New Year's Day of 1983, the *Teffé* was overflown by a Hercules C-130, followed by two F4 Phantom fighter planes loaded with missiles and by a Wessex anti-submarine helicopter. The following day, they were followed

[93] They were Adalbert Klaus Kolpatizk and Peter Joseph Barry (Martins, 1998, p. 36).

[94] Feijó (1983). Information regarding the involvement of the Clube Andino Paulista can be retrieved from the website of the Brazilian Navy: https://www.marinha.mil.br/secirm/proantar-noticias-alpinista. Accessed on: 16 November 2020.

[95] Henriques (1984, p. 110).

[96] Ibid.

[97] Bacila (1985, pp. 60–61) and Bakker (2004, p. 97).

[98] The Drake Passage is the body of water that separates the American continent from Antarctica and in which the Pacific and Atlantic Oceans converge. Its crossing is considered one of the most treacherous voyages, with waves that can reach more than 10 m high.

by a task force fleet some 40 miles away and were once more closely overflown by two Westland Lynx helicopters.[99] The encounter was uneventful and no communication between the British force and the Brazilian vessel was made, but it was a reminder of the conflict that a few months earlier have developed in the close vicinity of the Antarctic region.

In the first days of 1983, Brazil finally reached Antarctica, visiting several locations as shown in Fig. 3.1. The *Teffé* crossed the Drake Passage, passing the 60° of south latitude, officially entering into the Antarctic and reaching, on 5 January 1983, the King George Island (*25 de Mayo/Vaterloo*). Following the coast of the island to its southern margin, the Brazilian vessel entered Admiralty Bay (*Lasserre*), where it anchored in front of the Polish station, Henryk Arctowski. Due to the meteorological conditions, the Brazilians had to wait more than 14 h to set foot on Antarctic soil, when they disembarked and were warmly received by the Polish working at the Arctowski station.[100] In addition to entertaining them with food and drinks, the Poles let them fly the Brazilian flag on the station mast, marking the first time the Brazilian flag was officially raised in Antarctica.[101] From there, they visited the Russian station Bellingshausen, and the Chilean station Eduardo Frei, both located on the same island; the Chilean station Arturo Prat, on Greenwich Island; and the English Faraday station,[102] the US Palmer station and the Argentine Brown station along the Gerlache Strait.[103]

After exploring the north-western coast of the Antarctic Peninsula, the *Teffé* headed north in direction of Punta Arenas, Tierra del Fuego, to be refuelled. On 18 January, when in the Beagle Channel in the vicinity of Ushuaia, it was approached by radio by Argentine officials who ordered

[99] The accounts of these events are based on a *Manchete*'s article: Expedição: no rumo da Antártida (1983).

[100] The accounts in the magazine *Manchete* seem to suggest that the Brazilian expedition set foot in Antarctica on 6 January (Expedição: Brasil finca pé, 1983). However, all other accounts accessed during the research date the first disembarkation of the Brazilians on the 5 January. The account presented here is mainly based on: Brasileiros desembarcam na Antártida (1983).

[101] There were previous display of the Brazilian flag in Antarctica, including Gerlache rise of the flag on the *Belgica's* main mast (see Sections 2.1 and 2.3), and other Brazilian visitors' display of the flag in the continent. However, this was the first time that the flag was being flown officially in the continent.

[102] Currently operated by Ukraine as Vernadsky Station.

[103] Henriques (1984, p. 112).

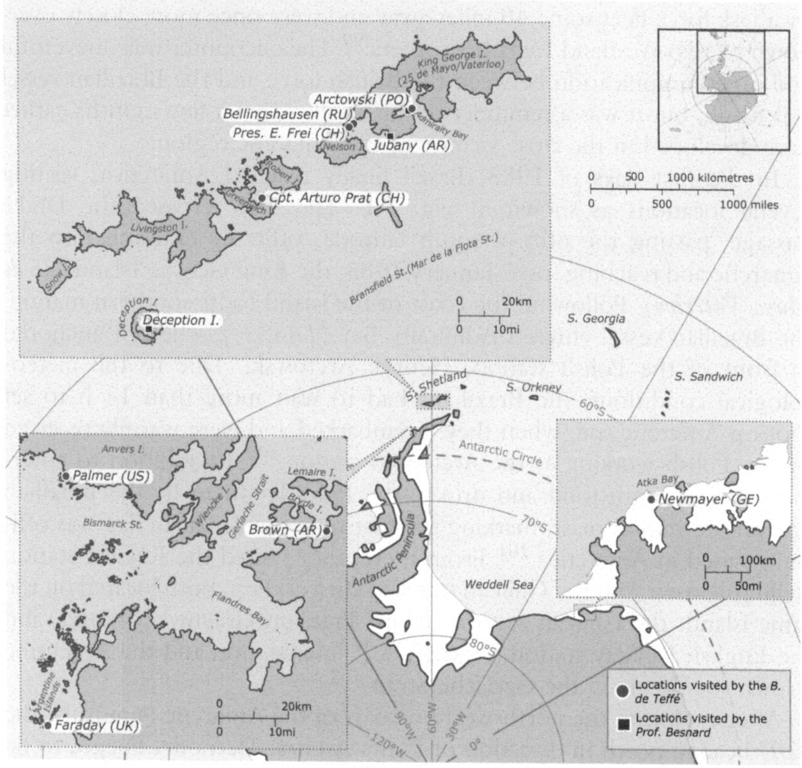

Fig. 3.1 Map of locations visited during the OPERANTAR I (*Notes* 1. The Argentine Jubany station was renamed Carlini in 2012. 2. The Faraday station complex was transferred to Ukraine in 1996 and renamed Vernadsky. *Source* The author)

it to stop to allow an Argentine pilot to embark, with the allegation that it was illegal to navigate the Channel in east–west direction without the presence of such official.[104] While the two helicopters were flying in the area, an Argentine coast guard vessel approached the *Teffé* with their anti-aircraft machineguns aimed at the helicopters. The Commander of the

[104] The accounts of these events are taken from: Barão de Teffé: 'Não attires (1983) and Henriques (1984, pp. 113–115).

Teffé opted not to accept the instructions and took a detour through the other margin of Navarino Island through the Murray Channel. The conflict did not finish there, and the Argentines intercepted the ship when it entered the Beagle Channel again, executing dangerous manoeuvres in the vicinity of the ship. In the end, the *Teffé* continued on its course unaltered to Punta Arenas without complying with the Argentine demands. The event prompted a protest from the Brazilian government, resulting in an official apology from the Argentines.[105] As with the event in the vicinity of the Malvinas/Falklands, it was another reminder of the disputes involved in the southern regions and the sensibilities of the Argentines.

After their stop in Punta Arenas, on 27 January, the *Teffé* headed in the south-easterly direction, towards the Weddell Sea, to reach the West German station Neumayer (see Fig. 3.1). In March 1982, the CIRM Secretariat Bakker had declared that Brazil intended to establish its Antarctic station on Princess Martha Land, on the East Region of the Weddell Sea, for which studying the region appeared as of fundamental importance.[106] However, the first expedition experience on navigating the area would prove it as a poor choice.

On 5 February, the *Teffé* crossed the Antarctic Polar Circle, reaching another milestone for Brazil. Though, due to mechanical problems, the trip had to be shortened and the vessel anchored at Atka Bay, before reaching the German station.[107] The visit to the station was possible thanks to the helicopters, that made about forty trips to move the personnel to the base. The incident served as a warning of what they could expect if they were willing to operate a base in that area. The mechanical problems of the ship made the expedition interrupt its programme and head back. On their way back to Brazil, new mechanical problems left the *Teffé* adrift for more than seventeen hours but, fortunately, the currents and prevailing winds took the ship north before control of the ship was regained. The ship left the Antarctic region on 12 February and reached

[105] Henriques (1984, p. 115).

[106] Almirante anuncia (1982). The region was the same defined by the projected Brazilian-Uruguaian expedition of the IBEA and Musso. In contrast to the assertions of Pericles Azambuja in his accounts of the events, for Bakker, the choice of such region was motivated by the fact that laid outside the sector claimed by Argentina, Chile and the United Kingdom (Bakker, 2004, p. 97).

[107] For this and the following account, see: Henriques (1984, pp. 115–116).

the port of Rio Grande on 22 February, docking in Rio de Janeiro six days after, completing the OPERANTAR I.[108]

The *Besnard* reached Antarctica on 9 January 1983 and immediately started its work in the Bransfield Strait (*Mar de la Flota*).[109] During its voyage, it visited Deception Island, the Argentine Jubany station[110] and the Polish Arctowski station (see Fig. 3.1), before heading to Punta Arenas for refuelling on 23 January. After four days in Punta Arenas, the ship resumed its work in the Bransfield Strait, extending it until 11 February, when it reached the port of Ushuaia. On the next day it headed back to Brazil. As the *Teffé* had experienced in their way south, it was overflown by the British Royal Air Force when close to the Malvinas/Falkland Islands. It arrived at Rio Grande on 19 February and at the port of Santos on 23 February.

While the *Teffé* and *Besnard* were in Antarctica, in Brazil, steps were taken to guarantee the future development of its Antarctic science. In January 1983, the CNPq announced special grants to be conceded for studies in the Geosciences, Atmospheric Sciences and Nature.[111] Whether the contacts with Antarctic researchers from other countries had driven the authorities to the realize that Brazilian scientists had the competences and capabilities for developing an Antarctic programme, Brazil wanted to train a group of scientists in polar specialities who could develop an academic career focused exclusively on Antarctica. In a decision that would eventually prove to be strategic, the CNPq offered some twenty grants. The first beneficiary of such a grant would be the young geologist Jefferson Cardia Simões, who would eventually play a key role in developing Antarctic science in Brazil.

Also in January, the geologist Antonio Rocha-Campos, coordinator of the management group of the PROANTAR, joined the Antarctic expedition of the US National Science Foundation, visiting the installations of the United States Antarctic programme in Christchurch, and the US McMurdo station and the New Zealand's Scott Base in Antarctica, before

[108] Nunes (2005, p. 118).

[109] Ibid., p. 113.

[110] Currently Carlini Station.

[111] Henriques (1984, p. 117).

reaching the South Pole and visiting the US Amundsen-Scott station.[112] The Rocha-Campos visit was an opportunity to acknowledge a complex programme as the one developed by the US, as well as the more modest programme developed by New Zealand and which could serve as a model for Brazil.

By the end of the 1982–1983 Antarctic season, Brazil had established its first contact with the frozen continent, and thus the tropic met the pole. But the event marked only the beginning of the involvement of the country in the region. To attain its objectives, Brazil needed to establish a consolidated scientific Antarctic programme, establish an Antarctic station, and become consultative member. At that time, those objectives still seemed far away, but some events would expedite that process.

3.3 Brazil's Definitive Hold in Antarctica

In the meantime, internationally, developments were taking place that would lead to the rapid accession of Brazil as a consultative member of the AT. The ATS was becoming subject to increasing criticism of considered as unequal, exclusivist and discriminatory. One of its critics, India, had a history of unsuccessfully trying to bring the question of Antarctica to the UN General Assembly in the years before to the AT, but gave it up after considering that the Treaty solved many of its concerns.[113] However, as discussions over natural resources advanced, that initial satisfaction with the provision of the Treaty turned into increasing criticism of its exclusivist nature.[114]

In fact, India's attitude towards Antarctica presented a strong parallel with Brazil in some respects. Its vision was imprinted with its intention to play a role in the international realm commensurate with its size and importance. Also, it had participated in the IGY but had not taken part of Antarctic activities. As well, it had refrained to make any claim while not renouncing to the possibility of doing so in the future. Likewise, it had

[112] Nunes (2005, pp. 116–117). Do not mistake the Scott Base of New Zealand at the Ross Island with the US Amundsen-Scott station, located on the geographic South Pole.

[113] Chaturvedi (2013, p. 313) and Dey (1991, p. 88).

[114] Actually, during the 1960s and early 1970s India's involvement in the Indo-China and Indo-Pakistan wars diverted its attention from Antarctica. Nevertheless, some Indian scientists participated as individuals in expeditions to Antarctica organized by other countries (Dey, 1991, p. 88).

established its interest in Antarctica mainly by the environmental relations between the continent and the meteorological phenomenon in India and the possibilities opened for developing countries by the Antarctic natural resources' exploitation prospects. Furthermore, the Department of Ocean Development was the main agency behind its first expeditions, something that resembled Brazil's decision to put its Antarctic endeavour under the coordination of the CIRM.

However, both countries policy departed from what they regarded as their attitude towards the Treaty since the middle of the 1970s. While Brazil accessed the Treaty in 1975 and since then pursued its inclusion in the AT discussions, India remained outside the agreement and pursued its own scientific activities in Antarctica without looking for approval from the AT members or coordinating with the SCAR. In December 1981, India launched its first Antarctic expedition, with twenty-one people on board of the icebreaker *Polar Circle*. The vessel, chartered to Norway, landed on 9 January 1982 in the so-called Norwegian sector, installing an unmanned solar powered weather station (*Dakshin Gangotri*) and returning to India in February of that year, after ten days in Antarctica.[115] A second expedition with twenty-eight people followed on the next season, and plans for establishing a permanent manned station started.

At the same time, the Prime Minister of Malaysia began to advocate the creation of a new Antarctic regime, presenting its proposal to the United Nations in September 1982 and in the Non-Aligned Summit Meeting in March 1983, in New Delhi.[116] The intention of Malaysia was to create a more open and egalitarian regime that could give voice to all countries without prior conditions such as the one established by the AT, seen as '...stringent qualifications...' by Malaysia.[117]

Both events, the actions of India in Antarctica outside the Antarctic regime—even when being completely directed to scientific aims and thus followed the basic spirit of the Treaty—and Malaysia's initiative, put pressure on the Antarctic regime at a moment when the decisive issue of mineral resources was being discussed, and thus, the regime was under close international scrutiny. Unintentionally, those developments benefitted Brazil.

[115] Beck (1983, p. 106).

[116] Ibid.

[117] Question of Antarctica (1984, p. 110).

In January 1983, an informal meeting took place along with the Special ATCM on Antarctic Mineral Resources (SATCM IV-2) in Wellington, New Zealand. Apart from other issues, the informal meeting discussed the Brazilian activities in Antarctica and evaluated the possibility of granting the country a consultative status.[118] Signalling a willingness to accept Brazil as a consultative member, the parties assigned Argentina and Chile the task to informally notify Brazil that a request to be incorporated with consultative status would be positively assessed by the parties if '…all minimum conditions were obliged.' The notification was given by the chief representative of the Chilean delegation, Ambassador Fernando Zegers, at a meeting on 4 February, before the Brazilian expedition was even back from Antarctica.[119] The communication revealed to the Brazilian government that access to consultative status was almost guaranteed and that the prerequisite of conducting 'substantial scientific activities' was somewhat relaxed. The pre-existent implicit condition of establishing a permanent station and a track record of continuous scientific research was substituted with the establishment of an ongoing long-term programme and plans for establishing a permanent station.[120]

On 20 April 1983, the President approved the regimentation of the CONANTAR through the Decree 88.245/83. The Decree incorporated new functions to the commission, including issues related to international cooperation, the political implications of guidelines proposed by the SCAR, the measures to be adopted as a consequence of the decisions taken at the ATCMs, the promotion of scientific events related to Antarctica, the adequation of the POLANTAR to the changes experienced by the Antarctic regime, and the assessment and approval of the proposals moved by the CIRM or a specific institution to be created. By the tone of the modifications, it was clear that Brazil expected to become a consultative member in the near future. Thus, the new functions adapted and complemented the prerogatives of the commission to facilitate the adaptation of Brazil to the normative and institutional setting derived from the Antarctic regime, including the installation of the CONANTAR as

[118] This and the following are based on: Antarctica (1983).

[119] During the research no information was found about if a similar notification was delivered by the Argentine government or if it was agreed between both countries that Chile would assume the approaching.

[120] Beck (1986, pp. 156–157).

an institutional interphase for the adaptation of national legislation to the decisions and regulations emanating from the international institution.[121]

By May 1983, the Brazilian government became aware of the intention of China to access the AT.[122] The Chinese intended to create a national polar research centre and establish its own station, looking, as Brazil and India were doing, to become a consultative party. China's government announcement was another factor playing in favour of Brazil's candidature, as, like India, it had only acceded to the Treaty at the last minute. Brazil's longest-term commitment with the Treaty was something that the country could boast in comparison with the other candidates. Also, China's candidacy helped to undermine Malaysia's project of promoting an alternative international arrangement, giving more legitimacy to an agreement that could not be considered as representing a small portion of the globe's population any longer.

In June 1983, the Secretary of the CIRM, Bakker, participated in a symposium about Antarctica in Kiel, Germany, and the next month, by invitation of the British government, visited the installations of the British Antarctic Survey, accompanied by Commander Neiva and Rocha-Campos. Continuing with the policy of gaining Brazilian favour, in that opportunity, the British offered to transfer a recently deactivated station to Brazil. Nonetheless, the offer was kindly declined by the Brazilians to avoid any Argentine outrage, due to the recent dispute of that country with the British, and the long-standing dispute over the Antarctic.[123]

After the return of the expedition, it was time to discuss the outcomes of the first expedition and to prepare for the next. The *Teffé* went through a series of adaptations to make it more suited for research and logistic activities while the *Prof. Besnard* was prepared for research as part of the BIOMASS.[124] One key issue was the determination of the location for the Brazilian station. Despite some favouring the installation on the sector defined by the frontage theory as belonging to Brazil,[125] the place chosen was selected due to practical reasons, included accessibility,

[121] This is not to say that it actually operated as intermediary, but that the Decree seems to be oriented in that direction.

[122] Informação N°263/83 (1983, May 25).

[123] Bakker (2004, p. 99).

[124] Question of Antarctica (1984, p. 6).

[125] See, for example: Henriques (1984) and Menezes (1982).

natural protection from natural elements, terrain characteristics and access to fresh water. Thanks to the experience gained in the first expedition, it was clear to the Brazilian Navy that they did not have the equipment or the experience and training necessary to operate in the Weddell Sea area, so they would need to limit to the Antarctic peninsula.[126]

In the speech of the Minister of Foreign Affairs, Ambassador Ramiro Saraiva Guerreiro, given at the opening of the symposium '*O Brasil na Antártida*', in August 1983, he relativized the convenience of claiming a region and concentrating the efforts within it, as it would result in the renunciation of all the remaining space.[127] As the interest in the potential of natural resource exploitation became more relevant, the territorial approach that characterized some circles of the Army contrasted with the more flexible and practical attitude of the Navy and Itamaraty, who aimed for some form of participation in any international scheme that could take advantage of Antarctic resources.[128]

In order to test the capabilities of air transportation to the region chosen for the station, a Lockheed C-130 Hercules of the Transport Squadron '*Gordo*' made the debut Antarctic flight for the Brazilian Air Force, landing on the Chilean airfield on the Marsh station on 23 August 1983. The flight not only established new possibilities for the Brazilian operations in the region, but also marked the inclusion of another Brazilian armed force within its Antarctic operations. In total, the Air Force made three flight missions during the 1983–1984 season to transport equipment and personnel between the Chilean airfields of Punta Arenas and Teniente Marsh.[129]

As response to the growing criticism of the closed and exclusive nature of the regime, Brazil, along with the other non-consultative parties of the AT, was invited as observer on the XII ATCM that was to be held

[126] Martins (1998, pp. 24–25).

[127] Saraiva Guerreiro (1983a, p. 6).

[128] An interest contrast between the defence of limiting the Brazilian actions to the sector defined by the frontage theory and the more sensible view focused on practical and scientific aspects that was rising within the Navy and Itamaraty could be identify in the interchange between Vice-Admiral Ibsen de Gusmão Câmara and Colonel Elber de Mello Henriques in the debate published in: A Antártica – Interesses Científicos e Econômicos (1982).

[129] Vásquez de Aquino (1985, pp. 85–86).

in Canberra in September.[130] On 24 August 1983, following the positive signal given to Brazil by the Chilean delegate on early February, the Brazilian government submitted a letter, accompanied by a report on 'Brazilian Antarctic Activities,' requesting the consideration of the consultative parties to the merits of Brazil to access consultative status. The favourable signal of the consultative parties was reinforced by the reflection of other international factors that favoured Brazil's candidacy. Malaysia's request to the UN Secretary-General to include the question of Antarctica in the Agenda—made on 11 August jointly with Antigua and Barbuda—pushed the consultative parties to display a greater openness and inclusiveness.[131] The same month and following the same expectation, India acceded to the Treaty and requested its consideration to become a consultative party. To facilitate a positive outcome, the Brazilian delegations start to make contact with representatives of the governments of the consultative parties, receiving mostly favourable prospects.[132]

As it was expected, on 12 September 1983, during a SATCM which preceded the XII ATCM, India and Brazil were accepted as the 16th and 17th consultative parties of the AT.[133] Their inclusion as consultative parties was a strategic move for the regime, as it shielded it from the criticism that Malaysia and other countries were propelling within the UN, and dissolved the prospects for any alternative arrangement. By including these countries, the Antarctic Treaty regime gained wider recognition from the developing world and blocked a unified front from the Non-Aligned Movement.[134] By a mixture of its own efforts and the opportunities provided by international circumstances, Brazil finally reached its long-coveted aim of having a say in the future of Antarctica. Nonetheless, it still had a long way to go to consolidate its position.

In the inaugural speech of the Consultative Meeting, the Brazilian head of the delegation, Ambassador Marcos Castrioto de Azambuja, thanked

[130] Beck (1986, p. 224).

[131] Saraiva Guerreiro (1983b, p. 229).

[132] See, for example: De embaixada em paris (1983), De embaixada em Moscou (1983), De Embaixada em Bruxelas (1983), De Embaixada em Toquio (1983), and De Embaixada em Londres (1983).

[133] They were jointly the 3rd/4th Party to be accepted after the original twelve consultative members, following the incorporation of Poland (1977) and West Germany (1981).

[134] Beck (1986, p. 224).

the distinction made by the other parties in recognizing Brazil as a consultative member, assuring them that the country had established a long-term Antarctic scientific programme and that:

> In exercising the responsibilities which it has taken as a Consultative Member, the Federative Republic of Brazil will strictly abide by the principles of peace, cooperation and freedom of scientific investigation contained in the Antarctic Treaty and established by the decisions and Recommendations which, in furtherance of those principles, have been adopted over a period of more than twenty years.[135]

Additionally, advancing what would constitute one of the pillars of the Brazilian attitude towards the continent, the Ambassador highlighted the importance of the protection of the Antarctic environment, noting that '...my Government has noted with satisfaction that the protection of the unique and fragile environment of Antarctica has been one of the main concerns of the Consultative Parties'.[136]

In terms of initiatives, Brazil only presented a joint proposal with Argentina and Chile that recommended an improvement in communication regarding meteorological data. The idea was to increase the reliability of weather forecasting and to favour a better distribution of weather stations, so as to have a more synoptical data-collection.[137] The main justification was the importance that the region had in the meteorological phenomena that occurred in the southern hemisphere, and thus was linked to the usual claim by Argentina, Brazil and Chile of the strong connection between South America and Antarctica.[138]

The achievement of consultative status was not the end of Brazil's initiative in Antarctica. To consolidate the position gained and to increase its Antarctic profile, one of the main objectives for the future of the programme was the installation of a Brazilian Antarctic station. The original plans for the next expedition envisaged the establishment of a

[135] Opening Address: Brazil (1983, p. 63).

[136] Ibid.

[137] Whether the final wording is different, and the general recommendation is more specific, it is likely that the proposal exerted influence on the issuing of Recommendation XII-1, on the collection and distribution of meteorological data, passed during that RCTA.

[138] For some details about the identity established in Argentina and Chile on that regard, see: Cardone (2020).

campsite in a suitable place for the preliminary study of the physical conditions that would allow the future installation of the station.[139] The scientific work would be carried out in neighbouring stations and in the camp itself. However, the camping plan evolved to a project for a modular facility that would allow a permanent presence. Thus, by mid-May, the idea of establishing a station had been adopted by Minister Fonseca, and the task was left to the CIRM Secretariat to be executed in the following six months.

The Secretariat of the CIRM established some basic conditions, including the possibility of future expansion, flexibility for instalment, the need for the facility to be transported on a single trip on board the *Teffé* and to favour national materials and technologies. Inspired in the West German station Neumayer, the solution adopted was a modular structure based on containers for ease of transportation and installation and which would keep costs low.[140] The contract was given to SAEF (*Sociedade Anonima de Equipamentos Ferroviários*) from São Paulo at the end of July 1983. The eight initial modules were designed and built in Brazil, taking six months to be delivered.[141] In addition, a special raft was built by the Navy to ferry the modules from the *Teffé* to the shore.

In the international political arena, Brazil's inclusion as consultative member had the immediate effect of positioning it in favour of the ATS. During the first round of debates in the UN General Assembly in 1983, the Brazilian representative, Mr. Souza e Silva, defended the regime, saying that:

> The Antarctic Treaty embodies the essential idea that freedom of scientific investigation in that continent and the exploration of its natural resources for the benefit of mankind can be fostered only through a sound and flexible system of co-operation based on mutual trust and general acceptance of the fundamental principles of the United Nations Charter. It has thus

[139] This and the following accounts of the process of planning and installation of the station are based—unless explicitly stated differently—on the excellent and detailed account of Martins (1998).

[140] Bakker (2004, p. 100).

[141] The chronology presented here is based on Martins (1998). Nunes (2005, p. 120) describes the timeframe as taking nine months, which included the bidding process. For how it was presented in the press and elsewhere, it seems that the PROANTAR gave great relevance to the fact of using domestic work, technologies and materials.

been successful in keeping Antarctica free of conflicts which throughout recent decades have disrupted international relations.[142]

The Brazilian defence of the Treaty in the General Assembly aimed to make its new role as consultative member compatible with its place within the Third World movement, becoming the voice of the latter inside the regime. In this regard, Brazil's intervention tried to ensure that the regime was not as closed as it appeared at first sight, and welcomed participation from experts from developing countries on its own activities, reassuring once more its position of being '...a developing country with a genuine and concrete interest in Antarctica.'[143]

As a recognition of the role that the state of Rio Grande do Sul played in promoting Brazil's involvement in Antarctica and the symbolism of being the southernmost state of Brazil, on 2 December 1983, the *Estação de Apoio Antártico* (Antarctic Activities Support Station—ESANTAR) was inaugurated in the campus of the *Fundação Universidade Federal do Rio Grande* (Federal University of Rio Grande) after ten months of work. The facility was the first fixed structure exclusively dedicated to the PROANTAR and it was in charge of the classifying, storing, fitting and distribution of the polar and scientific equipment of the programme.[144] At that time, the ESANTAR was also projected to develop as an international centre for Antarctic research due to its proximity to the port of Rio Grande, which was expected to become an international hub for Antarctic logistics, due to its better access and installations compared to the ports of Punta Arenas, in Chile, and Ushuaia, in Argentina.[145]

On the second Brazilian Antarctic expedition (OPERANTAR II), the *Besnard* left first, as the *Teffé* needed to wait until the modules for the station were ready. The *Besnard*left the port of Santos on 26 December 1983, to join the work being carried out by the Second BIOMASS Experiment in the Bransfield Strait (*Mar de la Flota*), as agreed a year before and making use of the experience gained in the previous season.[146]

[142] Summary Record of the 43rd Meeting (1983).

[143] Ibid.

[144] *Brasil na Antártica: 25 anos de história* (2008, p. 17).

[145] It was even considered the possibility of declaring the port of Rio Grande as a free port for Antarctic research vessels, as to promote its use. See: Bakker (2004, p. 98).

[146] Bakker (2004, p. 100).

After the departure of the *Besnard,* on 30 December, the *Teffé* received another visit of President Figueiredo, Minister Fonseca and the Minister of Foreign Affairs, Saraiva Guerreiro, to make a posthumous honorary recognition of Commander Ferraz—after whom the Brazilian station would be named—and launch the expedition, which set sail on 3 January 1984.[147]

The *Teffé* operation was divided into two phases. The first phase was to transport the modules and the personnel in charge of the installation of the base to the place of choice.[148] Whether priority was given to the recognition of several locations south of the Gerlache Strait (see Fig. 3.2), the expedition had the liberty to choose any other point considered as more suited. After leaving some scientists who would conduct their work at the Arctowski station thanks to the offer of Poland, the *Teffé* set course to west of the Antarctic Peninsula.[149] The locations on the Palmer Archipelago[150] proved to be very difficult for the operations, due to the presence of numerous icebergs and the strong winds.[151] Andvord Bay, on the western coast of the Antarctic Peninsula, offered good shelter from the bad weather of the Antarctic, but the presence of pack ice inside the bay suggested that its access could be closed for most of the year, including summer. With no better prospects south and little time to start the work of installation if they were to make it operative that season, the team in charge of the *Teffé* decided to go back north.

On 25 January 1984, the *Teffé* anchored in Admiralty Bay (*Lasserre*), King George Island (*25 de Mayo/Vaterloo*), and a team landed to explore the coast where the station was to be installed. The place of choice was the so-called British Point on the Keller Peninsula (see Fig. 3.2), location of an old whaling station and a derelict British meteorological base (Station G), some five miles, by sea, from the Arctowski station. This location was convenient as it offered shorter travel times and economy of resources, relative ease of access, a bay with good shelter, two sources of fresh water

[147] Nunes (2005, pp. 120–121).

[148] Bakker (2004, p. 99).

[149] Martins (1998, p. 49).

[150] The locations were: Dorian Bay (Wiencke Is.), Py Point (Doumer Is.) and Port Lockroy (Wiencke Is.). See: Bakker (2004, p. 100), Martins (1998, p. 49) and Nunes (2005, p. 121). Martins misplaces Py Point as being in Wiencke Is.

[151] This and the following is based on Martins (1998, pp. 51–54).

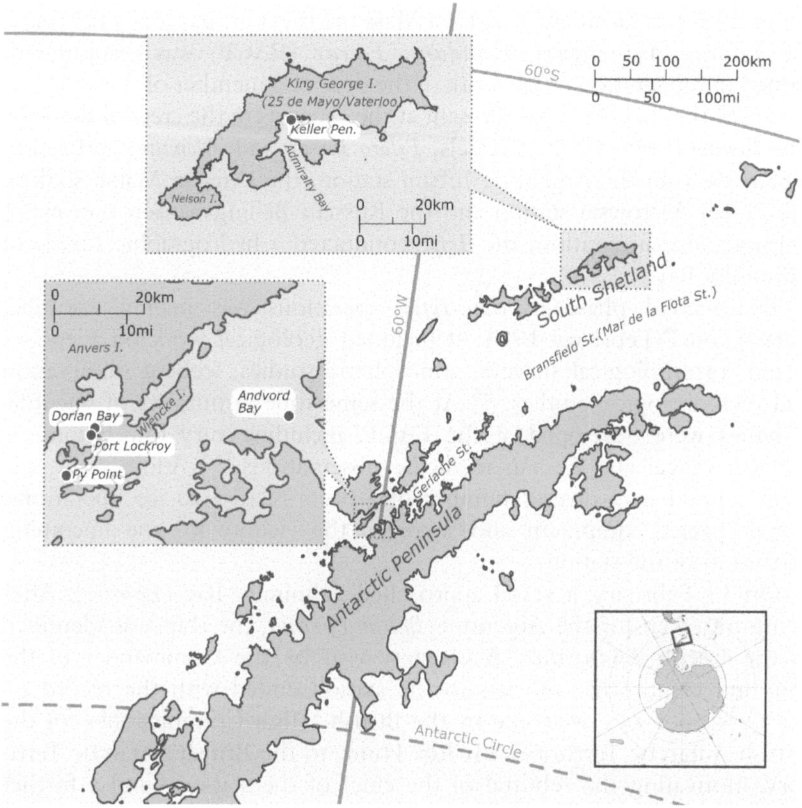

Fig. 3.2 Map of locations studied for the installation of the Brazilian Antarctic Station (*Source* The author)

and had many stations from other countries in close by, which facilitated mutual assistance. Also, the location allowed the team to take advantage of air operations through the Chilean Marsh airfield and ensured that the installation of the base could be done in a shorter timeframe.

That same day, the installation of the first Antarctic station of Brazil started, taking less than two weeks to be completed.[152] The facilities covered about 150 square metres, including two shared dorms and

[152] Martins (1998, p. 56) and Mattos (2015, p. 132).

common-room facilities.[153] After a Mass the previous day, on 6 February, the *Estação Antártica Comandante Ferraz* (EACF) was inaugurated, named in recognition of the work of the deceased member of the subcommittee of the PROANTAR. Present at the event were the crew of the *Teffé* and *Besnard*; two Chilean vessels, *Piloto Pardo* and *Alcazar*; and representatives from the Argentine Jubani station, the Chilean Marsh station, the Polish Arctowski station and the Russian Bellingshausen station.[154] During the inauguration, the *Teffé* conducted a hydrographic survey of Admiralty Bay.

The second phase of the *Teffé* operations was mainly scientific, starting on 7 February 1984. It included geological work on Elephant Island, ornithological surveys, atmospheric studies, iceberg observation and meteorological studies.[155] At the same time, a number of scientific activities were developed at the EACF, including very low frequency, meteorological and human medicine observations.[156] Additionally, the team trained in a diverse number of aspects related to ice operations, tested diverse equipment and surveyed the vicinity for the upcoming expansion of the station.

On 14 February, a vessel approached Admiralty Bay (*Laserre*). After being mistaken for the Argentine *Bahía Paraíso*, the ship was identified as the British *Endurance*. A courtesy visit by the Commander of the ship and some other officials to the station ended with the record on the guestbook of a welcome by the British 'High Commissioner for the British Antarctic Territory', Mr Rex Hunt, to the British Antarctic Territory, motivating the rebuttal of the chief of the station—and a further protest by the Brazilian government.[157] To repay the courtesy, a group of four visited the *Endurance* for a cocktail party. The visit of the British was followed, two days later, by the visit of the US vessel *Hero*, which was uneventful.

The *Teffé* returned to the EACF station on 6 March. Without conditions to remain open during the winter season, the team started the process of closing up and preparing the station for the next season. On

[153] Bacila (1985, p. 61).

[154] Nunes (2005, p. 122).

[155] Bakker (2004, p. 100).

[156] Martins (1998, p. 64).

[157] Martins (1998, pp. 69–71).

9 March, once the personnel at the Arctowski station had withdrawn and the EACF closed, the *Teffé* departed for Brazil, finishing its trip and docking in the port of Rio de Janeiro on 3 April 1984. Despite the great feat of the expedition, it was not received with as much fuss as one would expect for the installation of the first Brazilian Antarctic station.[158] The resignation, in late March, of Minister Maximiano da Fonseca due to a misunderstanding with the President could have been one of the reasons for the indifference of the achievements of the expedition, as Fonseca was the main promoter of the project.

While the OPERANTAR II was taking place, Brazil was continued in its role as consultative party, joining the discussion on the fourth formal round of the SATCM on Antarctic Mineral Resources that took place in Washington D.C. between 18 and 27 January 1984.[159] It also participated in late May in the fifth round that took place in Tokyo and in which Brazil offered to host the next meeting.

With the station in place, Brazil fulfilled most of what was considered to be the condition of a full member of the Antarctic regime. However, it still lacked membership of the SCAR, the main organization coordinating the scientific work in Antarctica. With two expeditions already completed and participation in the BIOMASS programme, Brazil certainly had the necessary credentials to access the scientific body. As the membership of the SCAR was in the name of National Scientific Committees, Brazil created, on 14 March 1984, the *Comité Nacional de Pesquisas Antárticas* (National Committee on Antarctic Research, CNPA until 1996),[160] dependent on the CNPq.[161] The Committee added to the institutional design was oriented to constitute a forum for debate on the main technological and scientific strategies to be submitted to the CONANTAR and the CIRM for consideration. Repeating the centrality of the Navy in the institutional scheme, its coordination was assigned to the CIRM Secretariat, having also representatives from the scientific community, the

[158] Martins (1998, p. 88).

[159] This and the following are based on: Saraiva Guerreiro (1984).

[160] The Committee will be recreated by Decree in 1996, receiving the acronym of CONAPA. It is probable that the creation of the Committee resulted in the withdrawal of the idea of any official Antarctic Institute in the molds of Argentina or Chile.

[161] Antártida. SCAR (1984).

Ministry of Foreign Affairs and the Ministry of Mines and Energy.[162] After the request, on 1 October 1984, during the XVIII plenary meeting of the SCAR in Bremerhaven, Brazil was unanimously accepted as a member jointly with India.[163]

Added to that, the institutional setting was modified. The internal organization of the subcommittee of the POLANTAR within the CIRM experienced some relevant changes. Up until then, the subcommittee met in plenary sessions, including all five coordinators of the five subprogrammes.[164] That scheme proved to be impractical, as not all the issues involved the expertise of every member and was difficult to match the agendas of all of them, as many came from different parts of the country.[165] Following these considerations, two groups were created, the Management Group (*Grupo de Gerenciamento*), which would be in charge of the scientific aspects of the programme, and the Operational Group (*Grupo de Operações*), in charge of the logistical and operational sides.[166]

Likewise, at the end 1984, Itamaraty experienced a general reorganization of the ministry that affected the responsibility of Antarctic issues.[167] The newly created *Divisão do Mar, da Antártica e do Espaço* (Seas, Antarctic and Space Division—DMAE), dependent from the Department of International Organizations, took over the Antarctic question, up until then under the responsibility of the Division of South America and the Division of the United Nations.[168] The conjunction of the issues of the seas, Antarctica and the space signalled once more a strong link between Antarctica and the maritime space.

[162] Ibid.

[163] China applied for membership that same year, but the plenary decided to wait for the reports of the Chinese marine research programme and the establishment of its summer station in the 1984/1985 season to be sent to the National Committees for its consideration the next Executive meeting (see: Scientific Committee on Antarctc Research [SCAR], 1985, p. 66).

[164] As defined in the original scientific programme: Life Sciences; Geosciences; Atmospheric Sciences; Logistics; and Education and Training.

[165] Martins (1998, p. 23, footnote 1, and p. 89).

[166] Ibid.

[167] Castro and Castro (2009, pp. 34–40).

[168] Ferreira (2009, p. 145).

By that time, the first outcomes of Brazilian research activities in Antarctica started to appear. The hydrographic work undertaken during the second expedition to Admiralty Bay (Laserre) of King George Island (*25 de Mayo/Vaterloo*) allowed the publication, on 28 September 1984, of the first Brazilian Antarctic Nautical Chart. Whether limited to a relatively small region, the chart was important as it presented the first tangible outcome of the technical and scientific work of the Navy in Antarctica. Additionally, the chart presented an important symbolic value, as it provided to Brazil a first sense of belonging through the development of new geographical knowledge of the region.

Added to that, the increasing investment in the programme signalled the commitment of the country with Antarctica. The third Antarctic expedition of Brazil (OPERANTAR III) implied a new level of operations, extending for more than five months and involving two round trips from Brazil to Antarctica and the transportation of a great quantity of materials and personnel. Besides the scientific programme, the main objective of this expedition was the extension of the EACF, that would incorporate 30 modules to the existing eight[169] and the installation of two shelters on distant islands. With those developments, Brazil expected to be able to extend its scientific programme and to make it possible to transform the EACF in an all-year station.

The expedition was launched on 15 November 1984 and, once more, involved the *Teffé* and the *Besnard* as well as Brazilian Air Force assistance through its C-130. On their arrival at Admiralty Bay, on 2 December, the Brazilians confirmed that the EACF was covered by snow, as already warned by the personnel at the Polish station. For that reason, the first task was to remove the snow and check the facilities, which proved to be in perfect condition.[170] After eleven days of working on the refitting of the station, the EACF was reactivated on 13 December.

Apart from the additional containers for the station, two shelters were installed, one located on 'Stinker Point' on the Elephant Island—chosen because of the relevance of the island for the krill industry[171]—named

[169] Some authors refer to thirty containers (see, for example: Bacila, 1985, pp. 22–23; Martins, 1998, p. 92; Nunes, 2005, p. 128) added to the station, whereas 'Brasil na Antártica' (*Op. Cit.,* p. 28) refers to twenty-eight. Without any specific documentation to support one or the other version, it has been taken the more prevalent information.

[170] Antártica, A Última Fronteira (1985).

[171] Bacila (1985, p. 61).

Engenheiro Wiltgen, and the other in the Nelson Island, named *Astrônomo Cruls*.[172] Both shelters were equipped for fitting 4 to 6 people and would serve only as temporary accommodation for small groups of scientists conducting observations *in loco*.

Added to the works on the station and the shelters, the *Teffé* continued its hydrographical work at Admiralty Bay (*Lasserre*). The team also removed the waste present on the shore and the vicinity of the abandoned British meteorological station,[173] and an Air Force engineer conducted studies for the installation of an airfield in the vicinity of the EACF, which later proved unfeasible due to the environmental impact.

On 13 December, after the reactivation of the EACF, the *Teffé* set sail for Brazil to load more material and additional modules, while officials and scientists remained to continue their work. On their way back to the vicinity of the Falkland/Malvinas, the *Teffé* was intercepted by a British frigate that requested identification, a reminder of the conflict of a few years before.[174] On 23 December the *Teffé* arrived at Rio Grande in Brazil, departing on 29 December for Antarctica for the second phase of the expedition.

Once the works of extension of the station were finished, the EACF was able to shelter up to twenty-two people, offering more comfortable and adequate facilities, included dorms, a kitchen, laundry, library, separate toilets for men and women, laboratories and other support facilities.[175] These works allowed the establishment of a diverse number of scientists in the fields of marine biology, palaeontology, astrophysics and meteorology.[176] With the programme completed, the *Teffé* finally arrived at Rio de Janeiro in 20 April 1985.

Meanwhile, other countries were creating their imprint on the Antarctic. In December 1984, Uruguay had inaugurated its station,

[172] *Brasil na Antártica: 25 anos de história* (2008, p. 29). Wiltgen was the first president of the IBEA and Cruls the astronomer that observed the transit of Venus on board of the *Parnaíba*.

[173] Bacila (1985, p. 26).

[174] Antártica, A Última Fronteira (1985), 39′30".

[175] *Brasil na Antártica: 25 anos de história* (2008, p. 29).

[176] Bacila (1985, pp. 22–23).

Artigas, following a similar approach to that used for the EACF.[177] Just two months later, on 20 February 1985, the Chinese inaugurated its first Antarctic station, Great Wall, counting on the participation of the Brazilians, as well as people from the nearby stations of Argentina, Chile, Germany, Poland, Russia and Uruguay, as was customary in the region.[178] Uruguay and China's actions in Antarctica were framed by the increasing interest that the region attracted as the negotiations for a mining agreement advanced. Latin-America was not alien to this interest, and besides Uruguay, Ecuador and Peru were taking steps to join as consultative parties, with indications of support from Brazil.[179]

In Brazil, political developments were taking place. Tancredo Neves was elected President, being the first civilian in holding the office after the 1964 military coup. Neves was a conservative, considered moderate and with a conciliatory tone with the military. However, it would not take office due to its sudden death. In his place, the elected vice-president José Sarney assumed the Presidency. Being closely linked to the military, Sarney had a difficult task ahead. He had to consolidate the transition to democracy in the context of a lasting economic crisis and growing social dissatisfaction. Despite those concerns were at the top of the governmental priority, the involvement of Brazil in Antarctica was already structured in a way that would ensure its continuity and consolidation.

Between 26 February and 8 March 1985, the SATCM on Antarctic Mineral Resources (SATCM IV-6) took place in Rio de Janeiro. It was the first meeting that accepted acceding parties as observers, giving a definitive end to the closed nature of the negotiations on mineral resources.[180] Also it was a first opportunity for Brazil to position itself within the landscape of the political arena of the regime, once it had defined a position regarding what should be the main objectives of its interventions in

[177] Martins (1998, pp. 87–88). Uruguay accessed to the Treaty in 1979, being acknowledged as acceding state in 1980. Allegedly its station was established at a location coveted by the Chinese (see: Historia del Uruguay en la Antártida, 2004).

[178] Bacila (1985, p. 47).

[179] As an example, between 1984 and 1986, during official visits, five joint declarations between the Presidents of Brazil and Ecuador (one), Peru (one) and Uruguay (three) covered the Antarctic issue, from proposing to collaborate in the interchange of experiences, to support the application for consultative status and scientific cooperation. For a brief summary of such acts, see: Cardone (2015, pp. 77–82). It is important to highlight that those countries were included within the Brazilian thesis of the frontage theory.

[180] Beck (1986, p. 198).

the negotiation: a special consideration to the developing world and the protection of the Antarctic natural environment.

In August 1985, at its opening of the XIII session of the CONANTAR, the recently appointed Ministry of Foreign Affairs, Olavo Egydio Setubal, recalled the already known position of Brazil regarding Antarctica, highlighting the interest linked with the climatological influences, the potential for economic exploitation of mineral resources and the strategic interests. But he also added the need to consider the particular desires and interests of Brazil as a developing country and the need to raise those issues within the discussions of the Antarctic regime, as well as the importance of promoting the preservation of the Antarctic environment, due to potential impacts on the Brazilian primary economy.[181]

From 7 to 18 October, the XIII ATCM took place in Brussels, Belgium, with the active participation of Brazil along with the other consultative members, to which China and Uruguay were incorporated after a Special Consultative Meeting on the first day of the debates. In his opening speech, the Brazilian head of Delegation, Ambassador David Silveira da Mota Jr, emphasized the increased interest and criticism that the Antarctic regime was receiving, attributed to the secrecy that the deliberations of the regime previously held—which Brazil itself had criticized in the past. Nevertheless, the Ambassador also stressed the little interest shown by the international community until then. As well, he emphasized the increasing number of countries that were accessing to the Treaty and the challenges that such broadening of the Treaty would imply, considering a review of the operation of the treaty necessary. In that regard, the Minister guaranteed Brazil's contributions as a developing country with considerable interests in Antarctica.[182]

During the meeting, Brazil proposed, jointly with Belgium, the recommendation on 'The Operation of the Antarctic Treaty System,'[183] which looked for a reporting scheme that would include the Consultative Meetings, the recently created Commission for the Conservation of Antarctic Marine Living Resource, and the SCAR. The proposal resulted in the approval of Recommendation XIII-2 on 'Operational reports at

[181] Setúbal (1985).

[182] Opening Address: Brazil (1985).

[183] The Operation of the Antarctic Treaty (1985).

ATCM,'[184] responding to a need already acknowledged by the increasing activity and participation in the regime.

Such political presence within the regime was coupled with the material presence in the continent as Brazil was consolidating its position there. The OPERANTAR IV took place on the 1985–1986 season and marked the end of the instalment process and the beginning of the permanent contact between Brazil and Antarctica. The beginning of the operation was marked by the expansion of the station with another eight containers added to the main installations and 6 to be dedicated to fuel (one oil and five diesel), what would allow the station to operate during the whole year without refuelling. In addition, the shelter Padre Balduíno Rambo was established in the King George Island (*25 de Mayo/Vaterloo*) by a group of geologists of the UNISINOS, thanks to the financial support of companies of Rio Grande do Sul.[185]

In March 1986, the EACF was visited by the Ministers of Foreign Affairs, the Navy and Science and Technology.[186] The official visit was an acknowledgement of the achievements of the programme and the start of a whole new phase in the Brazilian presence in the continent. For the first time, eleven people would endure the Antarctic winter in the expanded EACF station. Despite other officials having wintered in other countries' stations in previous seasons, this was the first time that the Brazilian presence was established on a continuous basis in a Brazilian facility, that is, on its own terms. Thus, on 2 March 1986, Brazil inaugurated its permanent presence in Antarctica. The tropic and the pole were definitively connected.

3.4 Conclusions

Although Antarctica was seen as a potential source of economic resources that could help foster the development of the country, the interest in the continent fell well behind other priorities, and the perspective adopted was pragmatic in political and economic terms. The influence of the frontage theory was felt in the political position adopted, but did not prevent the Brazilian government from fully accepting the international

[184] Operation of the Antarctic Treaty System (1985). Overview.

[185] Alvarez (1995, p. 124) and Alvarez and Casagrande (2005).

[186] Brasil (1986, p. 71).

regime nor conditioned the choices that the Antarctic programme of the country took, especially in what concerned the location of the Brazilian station. Also, the more immediate interest in the Itaipu dam project prioritized relations with Argentina, delaying for several years the effective organization of an official expedition, until the dispute was mostly settled, and the Argentines already declared their support for the Brazilian Antarctic involvement.

With the clear objective of gaining access to the political decision-making process of the Antarctic regime, the Brazilian policy for Antarctica oriented its institutional design to be of minimal scale, flexible and demanding little resources. That allowed a rapid organization of the scheme and the concentration of resources in the practical and operational aspects of the project. However, it required the pressure of the private initiatives and the public opinion and the willingness of the Navy to finally take the steps for the first official Antarctic expedition.

Without the Navy, and specially without Minister Fonseca, it is unlikely that the Antarctic expedition would have taken place in 1982, especially taking into consideration the acute economic crisis that the military regime was going through. In that sense, the view of the maritime space as a potential source for natural resources that could refuel national development was key in pushing the Antarctic endeavour through the interest expressed in the potential of oil exploitation and the negotiations on Antarctic mineral resources—it is necessary to note that the same interest was not expressed in Antarctic maritime living resources. That helped to explain both the centrality that the Navy had in the original institutional design and its future evolution, as well as the maritime character that the Brazilian Antarctic programme acquired.

The original scheme in which different military branches and civil departments shared a wide range of operational and research responsibilities ended up in a clear predominance of the Navy, followed by the scientific sector, with the limited participation of the Air Force to provide air support. The originally designed Brazilian Antarctic Institute never materialized and the CIRM assumed most of its responsibilities.

That maritime character was not only expressed in the Antarctic programme per se, but it was reflected also in the place that the question had in Itamaraty. After a first assignment with the international organizations section, the inclusion of Antarctica along with the issues of sea and outer space also revealed a tendency to consider Antarctica on parallel with the maritime question.

That characteristic was expressed in the connexions that Brazil established with the Antarctic region, expressed in economic terms. As the blue Amazon presented the opportunity to benefit from the natural resources from the sea-bed, the possibilities of oil exploitation in Antarctica made it necessary for Brazil to require participation in its exploitation under conditions that took into account the different levels of technological development. No less significantly, the Antarctic was seen to be connected to the national territory economy through the environmental connections, established through the maritime currents and the atmospheric dynamics, which could alter the harvesting of maritime natural resources and the profits deriving from the primary activities in the national territory.

This conditioned the vision that the country had for its Antarctic scientific programme, which also adopted a pragmatic character. In general, all scientific lines of research were established with a special view to potential economic direct or indirect benefits. In that regard, this interest was not placed in the development of a particular field of science nor in the knowledge of the Antarctic region, but in the possible benefits that such knowledge could bring to national development.

Of course, this approach was not monolithic nor straightforward, and researchers encountered a wide degree of freedom in proposing their research. However, since the inception of the first policy, it imprinted a strong instrumental role and a close link with the economic interest.

References

Alvarez, C. E. de. (1995). *Arquitetura na Antártica: Ênfase nas edificações brasileiras em madeira*. University of São Paulo. http://lpp.ufes.br/arquitetura-na-antártica-ênfase-nas-edificações-brasileiras-em-madeira

Azambuja, P. (2005). *O sonho do aurora austral: como o Brasil chegou à Antártida*. Magna Quies.

Bacila, M. (1985). *Cartas da Antártica: Estação Antártica Comandante Ferraz*. Imprensa da Universidade Federal de Parana.

Bakker, M. P. R. de. (2004). O início da participação brasileira no continente antártico - algumas considerações sobre os recursos do mar: um depoimento. *Revista Marítima Brasileira*, *124*(4–6), 83–106. http://eprints.uanl.mx/5481/1/1020149995.PDF

Beck, P. J. (1983). India in Antarctica: Science -and politics- on ice. *Nature, 306*, 106–107.

Beck, P. J. (1986). *The international politics of Antarctica*. Routledge Revivals. http://eprints.kingston.ac.uk/25053/

Brasil, M. das R. E. do. (1986). *Ministério das Relações Exteriores - Relatório 1986*. Brasília.

Brasil na Antártica: 25 anos de história. (2008).

Cardone, I. J. (2015). *As Posições Brasileiras No Sistema Do Tratado Antártico Com Ênfase Na Questão Ambiental*. Universidade Federal de Parana. http://hdl.handle.net/1884/38844

Cardone, I. J. (2020). Shaping an Antarctic Identity in Argentina and Chile. *Defence Strategic Communications*, *8*, 53–88. https://doi.org/10.30966/2018.RIGA.8.2

Castro, F. M. de O., & Castro, F. M. de O. (2009). *Dois Séculos de História da Organização do Itamaraty: 1808–2008 - Vol 2: 1979–2008* (Vol. 2). Fundação Alexandre de Gusmão.

Chaturvedi, S. (2013). Rise and decline of Antarctica in Nehru's geopolitical vision: Challenges and opportunities of the 1950s. *The Polar Journal*, *3*(2), 301–315. https://doi.org/10.1080/2154896X.2013.868087

Dey, A. (1991). India in Antarctica: Perspectives, Programmes, and Achievements. *Polar Record*, *27*(161), 87–92. https://doi.org/10.1177/0020881792029002003

El-Sayed, S. Z. (1991). The contribution of the BIOMASS program to antarctic marine ecosystem research. In G. di Prisco, B. Maresca, & B. Tota (Eds.), *Biology of antarctic fish* (pp. 131–142). Springer-Verlag. https://link.springer.com/content/pdf/10.1007%2F978-3-642-76217-8_9.pdf

Ferreira, F. R. G. (2009). *O Sistema do Tratado da Antártica: evolução do regime e seu impacto na política externa brasileira*. Fundação Alexandre de Gusmão.

Fontes Diégues, F. M. (1987). Cinco anos de presença da Marinha na Antártida. *Revista Marítima Brasileira*, *107*(10–12), 97–106.

Fontes, W. (2012). *Los aportes del Profesor Julio C . Musso al trabajo pionero del Instituto Brasileño de Estudios Antárticos (IBEA)*. Pichidangui.

Great Britain. (1981). *The United Kingdom defence programme: The way forward*. London.

Henriques, E. de M. (1984). *Uma Visão da Antártica*. Biblioteca do Exército Editora.

Machowski, J. (1991). Poland's policies toward Antarctica. *German Yearbook of International Law*, *34*, 63–91.

Marinho, F. E. L. R. (2016). *A última fronteira: a construção de um território brasileiro no Atlântico Sul (1960–1979)*. Universidade Federal do Rio Grande do Norte.

Martins, E. N. (1998). *Retorno a Ferraz: histórias de um pioneiro*. Editora Universidade de Brasília.

Mattos, L. F. de. (2015). *O Brasil e a Adesão ao Tratado da Antártica: uma análise de política externa no governo Geisel*. Universidade Federal Fluminense.
Menezes, E. C. de. (1982). *Antartida e os Desafios do Futuro*. Capemi Editora.
Nunes, M. R. (2005). *O Brasil na Antártica*. Régis Aló.
Pinto Coelho, A. (1984). O momento antártico brasileiro. *Revista Marítima Brasileira, 104*(7–9), 81–90.
Saraiva Guerreiro, R. (1982). *Primeira Reuniao da Comissão Nacional para Assuntos Antárticos. Resenha de Política Exterior do Brasil*. http://www.itamaraty.gov.br/
SCAR. (1959). Special committee on Antarctic research bulletin n.3. *Polar Record, 9*(63), 589–608.
Scientific Committee on Antarctic Research (SCAR). (1978). Scientific Committee on Antarctic Research Bulletin No 60, September 1978—Fifteenth Meeting of SCAR, Chamonix, France, 16 to 26 May 1978. *Polar Record, 19*(120), 295–325. https://www.scar.org/scar-library/reports-and-bulletins/scar-bulletins/4320-scar-bulletin-60/
Scientific Committee on Antarctic Research (SCAR). (1985). Scientific Committee on Antarctic Research Bulletin No 79, January 1985—Eighteenth Meeting of SCAR, Bremerhaven, FRG, 1 to 5 October 1984. *Polar Record, 22*(139), 455–472. https://doi.org/10.1007/978-94-009-8261-1_44
Spektor, M. (2002). O Brasil e a Argentina entre a cordialidade oficial e o projeto de integração: A política externa do governo de Ernesto Geisel (1974–1979). *Revista Brasileira De Política Internacional, 45*(1), 117–145.
Vásquez de Aquino, S. T. (1985). A presença brasileira na Antártida. *Revista Marítima Brasileira, 105*(7–9), 77–89.
Vicuña, F. O., & Araya, A. S. (Eds.). (1977). *El desarrollo de la Antártica*. Editorial Universitaria—Instituto de Estudios Internacionales de la Universidad de Chile.

Archival Sources

A Antártica – Interesses Científicos e Econômicos do Brasil. (1982, July). Cadernos de Estudos Estratégicos, Issue 2 [Clipping]. Centro Brasileiro de Estudos Estratégicos – CEBRES, pp.13–27 (BR DFANBSB V8.MIC, GNC.CCC.87015378), Arquivos Nacionais, Brazil.
Abreu, H. A. (1976a, June 7). [Exposição de motivos DAM-I/015/210(D)] Divisão de Segurança e Informações do Ministério da Justiça - BR RJANRIO TT, pp. 52–56 – Exposição de Motivos 016/76 (BR/RJANRIO/TT/0/JUS/PRO/0538), Arquivo Nacional, Brazil.
Abreu, H. A. (1976b, October 28). [Exposição de Motivos 016/76]. Divisão de Segurança e Informações do Ministério da Justiça -

BR RJANRIO TT, pp. 83–85 – Exposição de Motivos 044/76 (BR/RJANRIO/TT/0/JUS/PRO/0538), Arquivo Nacional, Brazil.

Antarctica (1983, February 4). [Report from Brazilian Embassy in Santiago de Chile]. Antártida. – Dossiê, pp. 1027–1028 (BR DFANBSB Z4.REX.IPS.27), Arquivo Nacional, Brazil.

Antártida Argentina: ¿Qué pretenden los brasileños? (1976, February 25). [Clipping from an unidentified Argentine newspaper]. Antártida. – Dossiê (BR DFANBSB Z4.REX.IPS.27). Arquivo Nacional, Brazil.

Antártida. SCAR. Filiação do Brasil. (1984, August 24). [Information from the Embassy in Brussels]. Antártida. – Dossiê, pp. 988–990 (BR DFANBSB Z4.REX.IPS.27), Arquivo Nacional, Brazil.

Anteprojeto de política nacional para assuntos antárticos – Subsídios (ca. 1976). [Guidelines for Brazilian Policy for Antarctica] Divisão de Segurança e Informações do Ministério da Justiça - BR RJANRIO TT, pp. 57–71 – Documento N°1 (BR/RJANRIO/TT/0/JUS/PRO/0538), Arquivo Nacional, Brazil.

Araujo, J. H. P. de. (1977, January 14). [Correspondence with Dr Teles, President of the CNPq]. Antártida. – Dossiê, pp. 1055. Relações Brasileiro-Britânicas em assuntos antárticos (BR DFANBSB Z4.REX.IPS.27), Arquivo Nacional, Brazil.

Às Embaixadas em Washington, Buenos Aires, Santiago... (1975, June 3) [Política Antártica. Pedido de Subsídios]. Antártida. – Dossiê, pp. 740–741 (BR DFANBSB Z4.REX.IPS.26), Arquivo Nacional, Brazil.

Aspectos da Conjuntura Política Nacional – Análise N°3 (SG/CSN). (1982, March). [Report on national political context]. Política interna - aspectos da conjuntura política nacional, análises da IA. SC, pp. 24–36 (BR DFANBSB N8.0.PSN, EST.0423), Arquivo Nacional, Brazil.

Aspectos da Conjuntura Política Nacional – Análise N°5 (SG/CSN). (1982, September). [Report on national political context]. Política interna - aspectos da conjuntura política nacional, análises da IA. SC, pp. 62–71 (BR DFANBSB N8.0.PSN, EST.0423), Arquivo Nacional, Brazil.

Castro, J. A. de A. (1975, July 24). [Política Antártica, contato com o secretario do SCAR]. Antártida. – Dossiê, pp. 1134–1136 (BR DFANBSB Z4.REX.IPS.27), Arquivo Nacional, Brazil.

Comitê Científico para a Pesquisa Antártica (SCAR). (1975, September 19). [Report on the contacts made with the SCAR]. Antártida. – Dossiê, pp. 1111–1114 (BR DFANBSB Z4.REX.IPS.27), Arquivo Nacional, Brazil.

De Embaixada em Bruxelas para Exteriores. (1983, September 2). Antartida. Brasil. Categoria Consultiva no Tratado [Information from legation in Bruxels]. Antártida. – Dossiê, p. 1006 (BR DFANBSB Z4.REX.IPS.27), Arquivo Nacional, Brazil.

De Embaixada em Londres para Exteriores. (1983, September 6). Antartida. Brasil. Categoria Consultiva no Tratado. Reino Unido [Information from legation in London]Antártida. – Dossiê, p. 1003 (BR DFANBSB Z4.REX.IPS.27), Arquivo Nacional, Brazil.

De Embaixada em Moscou para Exteriores. (1983, September 2). Antartida. Brasil. Categoria Consultiva do Tratado [Information from legation in Moscow]. Antártida. – Dossiê, p. 1007 (BR DFANBSB Z4.REX.IPS.27), Arquivo Nacional, Brazil.

De Embaixada em Paris para Exteriores. (1983, September 2). Antartida. Brasil. Categoria Consultiva no Tratado [Information from legation in Paris]. Antártida. – Dossiê, p. 1010 (BR DFANBSB Z4.REX.IPS.27), Arquivo Nacional, Brazil.

De Embaixada em Toquio para Exteriores. (1983, September 6). Antartida. Brasil. Categoria Consultiva no Tratado. Posição Japonesa [Information from legation in Tokio]Antártida. – Dossiê, pp. 1004–1005 (BR DFANBSB Z4.REX.IPS.27), Arquivo Nacional, Brazil.

Grupo de Trabalho sobre Antártida. (1975, August 18). [Invitation to several departments]. Antártida. – Dossiê, pp. 1121–1130 (BR DFANBSB Z4.REX.IPS.27), Arquivo Nacional, Brazil.

Informação N°263/83. (1983, May 25). Atividades da RPC. Adesão ao Tratado da Antártida [Report from Brazilian Embassy in Beijing]. Antártida. – Dossiê, pp. 1017–1019 (BR DFANBSB Z4.REX.IPS.27), Arquivo Nacional, Brazil.

IX Reunião Consultiva do Tratado da Antártida. (1977, March 1). [Information regarding British support]. Antártida. – Dossiê, pp. 1051–1053. (BR DFANBSB Z4.REX.IPS.27), Arquivo Nacional, Brazil.

Krill and Other Marine Activities. (1976, November 9). [File 72/636]. (NAA/B1387/81/70). National Archives of Australia.

Pitão, M. C. de M. (1976, May 10). Memorandum para o Chefe do Departamento Econômico [Memorandum with Enclosed documents]. Antártida. – Dossiê, pp. 1077–1085. (BR DFANBSB Z4.REX.IPS.27), Arquivo Nacional, Brazil

Política Antártica Brasileira. (1975, July 16) [Information to the Brazilian Embassy in London]. Antártida. – Dossiê, pp. 1144–1146. (BR DFANBSB Z4.REX.IPS.27), Arquivo Nacional, Brazil

Política e Tecnologia, Internacional. Aproveitamento dos Recursos Minerais da Antártida. (1976, June 5). [Telegram]. Arquivo Antônio Azeredo da Silveira – Serie Chancelher N° 69 (AAS mre d 1974.03.26/DOC IX-I), CPDOC-FGV, Brazil.

Política Nacional para Assuntos Antárticos. (ca. 1976a). [Draft Brazilian Antarctic Policy]. Divisão de Segurança e Informações do Ministério da Justiça - BR RJANRIO TT, pp. 72–75 – Documento N°2 (BR/RJANRIO/TT/0/JUS/PRO/0538), Arquivo Nacional, Brazil.

Política Nacional para Assuntos Antárticos. (1976b, October 28). [Brazilian Antarctic Policy]. Divisão de Segurança e Informações do Ministério da Justiça

- BR RJANRIO TT, pp. 86–91 – Política Nacional para Assuntos Antárticos – Diretrizes Gerais (BR/RJANRIO/TT/0/JUS/PRO/0538), Arquivo Nacional, Brazil.

Relações Brasil-Austrália. Cooperação na Antártida. (1976, April 9). [Brasemb Canberra]. Antártida. – Dossiê, p. 1097 – DAM-I/DAOc. (BR DFANBSB Z4.REX.IPS.27), Arquivo Nacional, Brazil.

Saraiva Guerreiro, R. (1983a). Circular Postal N°7 737, DNU/210(D), 1983/2—Anexo: Palestra proferida por Sua Excelencia... [Transcript of discourse]. Divisão de Segurança e Informações do Ministério das Relações Exteriores, pp 361–369 (BR DFANBSB Z4.REX.PPR.18), Arquivos Nacionales, Brazil.

Saraiva Guerreiro, R. (1983b, August 17). Antártida. Brasil. Admissão ao nível de Parte Consultiva do Tratado [Information for the Presidency]. Arquivo Saraiva Guerreiro – Ministerio das Relações Exteriores, pp. 226–230 (SG mre 1979.03.20/DOC XXXII - DOC 231). CPDOC-FGV.

Saraiva Guerreuro, R. (1984, November 12). Antártida. Regime para exploração e aproveitamento de recursos minerais. Reunião no Rio de Janeiro [Information for the Presidency]. Antártida. – Dossiê, pp. 982–983 (BR DFANBSB Z4.REX.IPS.27), Arquivo Nacional, Brazil.

Silveira, A. F. A. (1975a, August 18). [Information for the Presidency]. Arquivo Antônio Azeredo da Silveira – Comitê Científico para a Pesquisa Antártica (SCAR) (AAS mre d 1974.03.26/DOC IX-30), CPDOC-FGV, Brazil.

Silveira, A. F. A. (1975b, July 8). [Information for the Presidency]. Arquivo Antônio Azeredo da Silveira – Instrumentação da política brasileira em assuntos antárticos. (AAS mre d 1974.03.26/DOC VIII-45), CPDOC-FGV, Brazil.

Silveira, A. F. A. (1976a, February 16). [Information for the Presidency]. Arquivo Antônio Azeredo da Silveira – Relações Brasil-Austrália. Ida de um cientista brasileiro... (AAS mre d 1974.03.26/DOC XIV-17), CPDOC-FGV, Brazil.

Silveira, A. F. A. (1976b, February 17). [Correspondence to Adm. Henning]. Antártida. – Dossiê, pp. 1098–1099 – Relações Brasil-Grã Bretanha. Antártida. (BR DFANBSB Z4.REX.IPS.27), Arquivo Nacional, Brazil.

Silveira, A. F. A. (1976c, February 16). [Information for the Presidency]. Arquivo Antônio Azeredo da Silveira – Cooperação brasileiro-britânica em assuntos antárticos. Convite... (AAS mre d 1974.03.26/DOC XXVIII-65), CPDOC-FGV, Brazil.

Silveira, A. F. A. (1977a, February 22). Reunião do Comitê Científico para Pesquisa Antártica (SCAR) [Information for the Presidency]. Arquivo Antônio Azeredo da Silveira – Reunião do Comitê Científico para Pesquisa Antártica (AAS mre d 1974.03.26/DOC XXXI-76), CPDOC-FGV, Brazil.

Silveira, A. F. A. (1977b, June 8). [Report to President Geisel]. Arquivo Antônio Azeredo da Silveira - DAM-I/167/210(D) (AAS mre d 1974.03.26/DOC XXXI-76/A1), CPDOC-FGV, Brazil.

Souza, C. G. (presumed). (1976a, October 22). Política Externa. Antártida. Declarações do "Diretor Nacional da Antártida" [Telegram from the Brazilian Embassy in Buenos Aires]. Antártida. - Dossiê, pp. 1067–1068 - DAM-I/ADITEL 3488. (BR DFANBSB Z4.REX.IPS.27), Arquivo Nacional, Brazil.

Souza, C. G. (presumed). (1976b, October 25). Política Externa. Antártida. XIV Reunião do SCAR [Telegram from the Brazilian Embassy in Buenos Aires]. Antártida. - Dossiê, pp. 1063–1066 - DAM- I/DE-11/DNU. (BR DFANBSB Z4.REX.IPS.27), Arquivo Nacional, Brazil.

Newspapers and Other Sources

Almirante anuncia que o Brasil pretende enviar missão à Sub-Antártida. (1982, March 18). *Jornal do Brasil* (Rio de Janeiro, Brazil), Year 91, Issue 340, 5.

Alvarez, C. E. & Casagrande, B. (2005). Resultado da Avaliação Ambiental de Eficiência da Técnica Construtiva e do Planejamento Logístico no Desmonte do Refúgio Padre Balduíno Rambo [Report]. DI/53, XVI RAPAL.

Antártica, A Última Fronteira [Antarctica. The last frontier] (1985). [Documentary]. Rede Manchete.

Barão de Teffé: 'Não attires em mim, Argentina.' (1983, February 5). *Manchete* (Rio de Janeiro, Brazil), Issue 1607, 14–16.

Berg, M. (1981, December 5). Brasil e Inglaterra brigam pelo navio polar. *Manchete* (Rio de Janeiro, Brazil), Issue 1546, 14–15.

Brasileiros desembarcam na Antártida com emoção (1983, January 6). *Jornal do Brasil* (Rio de Janeiro, Brazil), Year 92, Issue 271, 9.

Brasil prepara expedições à Antártida e Subantártida para início de dezembro (1982, July 3). *Jornal do Brasil* (Rio de Janeiro, Brazil), Year 92, Issue 86, 2.

Brasil prepara missão de reconhecimento à Antártida (1982, February 1). *Jornal do Brasil* (Rio de Janeiro, Brazil), Year 91, Issue 297, p.6.

Buque hidrográfico A.R.A. 'Islas Orcadas' 1974–1979. Source: https://www.histarmar.com.ar/Armada%20Argentina/Buques1900a1970/HidrOceanograficos/IslasOrcadas-.htm. Accessed on: 10 November 2020.

Decreto N° 86.829 (1982, January 12). Brazil.

Decreto N° 86.830 (1982, January 12). Brazil.

Decreto N° 88.245 (1983, April 20). Brazil.

II Plano Setorial para os Recursos do Mar, 7 April 1986. Decreto 92.522/86.

Expedição: Brasil finca pé (e Bandeira) na Antártida (1983, January 22). *Manchete* (Rio de Janeiro, Brazil), Issue 1605, 125.

Expedição: no rumo da Antártida, Malvinas (1983, January 15). *Manchete* (Rio de Janeiro, Brazil), Issue 1604, 120.

Feijó, A. (1983, January 1). Exploração: O Brasil na Antártida. *Manchete* (Rio de Janeiro, Brazil), Issue 1602, 1 January 1983, 121.

Final Report of the First Special Antarctic Treaty Consultative Meeting. (Ca. Late 1977). The Antarctic Treaty. Report of the First Special Consultative Meeting held at London 25, 17 and 19 July 1977 and Recommendations of the Ninth Consultative Meeting held at London 19 September – 7 October 1977. Antarctic Treaty Secretariat.

Krill/O Alimento do Futuro (1981, February 28). *Manchete,* (Rio de Janeiro, Brazil), Issue 1506, 144–146.

Opening Address: Brazil, H. E. Ambassador David Silveira da Mota Jr. (1985) – Antarctic Treaty. Final Report of the Thirteenth Antarctic Treaty Consultative Meeting, Brussels, 7–18 October 1985, pp. 125–128. Antarctic Treaty Secretariat.

Opening Address: Brazil, H.E. Ambassador Marcos C. de Azambuja (1983) - Antarctic Treaty, Report of the Twelfth Consultative Meeting, Canberra 13–27 September 1983, p. 63. Antarctic Treaty Secretariat.

Operation of the Antarctic Treaty System, Overview (1985). Final Report of the Thirteenth Antarctic Treaty Consultative Meeting, Brussels, 7–18 October 1985, pp. 34–35. Antarctic Treaty Secretariat.

Polônia reduz expedições à Antártida e pode ceder navio polar ao Brasil. (1981, December 4). *Jornal do Brasil* (Rio de Janeiro, Brazil), Year 91, Issue 240, p. 8.

Question of Antarctica (1984, November 2). AGNU A39/583/Part II, Vol. II, [English].

Setúbal, O. E. (1985, August 2). Discurso pronunciado por sua excelência o Senhor Ministro de Estado das Relações Exteriores, Doutor Olavo Egydio Setubal por ocasião da abertura da XIII Sessão da Comissão Nacional para Assuntos Antárticos (CONANTAR) [Transcript of speech]. (Colleção Olavo Setúval), Itaú Cultural, Brazil. Source: https://www.itaucultural.org.br/sites/olavo-setubal/ministro.html. Accessed on: 23 November 2020.

Summary Record of the 43rd Meeting (1983, November 29) – Question of Antarctica (Continued). (A/C.1/38/PV.43), United Nations.

Teixeira (1976, January 17). Cómo é rica a Antártida Brasileira. *Machete* (Rio de Janeiro, Brazil), Issue 1239, 120–121.

The Operation of the Antarctic Treaty System (Item 5) (1985). [Working Paper]. RCTA XIII/WP31. Antarctic Treaty Secretariat.

CHAPTER 4

The Rise of the Newcomers

4.1 Brazil's Political Duality

As the negotiations for the mining agreement advanced, the Antarctic Treaty System (ATS) attracted wider attention. The issue acquired even more prominence as, in 1985, scientists of the British Antarctic Survey discovered that, since the late 1970s, the ozone layer was depleting rapidly in the region above Antarctica, giving a dark panorama for the years to come.[1] Whether the issue of the ozone layer had no direct relationship with the prospects of mining activities in Antarctica, it did link the human action with its negative impacts over the Antarctic environment. Coupled with the concerns related to the potential impacts of mining activities there, the ozone layer depletion brought the Antarctic regime within the international agenda.[2]

The environmental question not only located Antarctica within the main issues of global concern but also resulted in the involvement of another kind of actor, the environmental activist groups.[3] Therefore, by the second half of the 1980s, the ATS started to suffer significant pressures, including the challenge posed by Malaysia and its allies, the request

[1] Farman et al. (1985).

[2] Villa (2004, Chapter 4).

[3] A specific analysis of this aspect could be seen in Villa, *Op. Cit.*

I. J. Cardone, *The Antarctic Politics of Brazil*,
https://doi.org/10.1007/978-3-030-80161-8_4

for increased openness and participation of new parties, and the involvement of non-governmental organizations looking to influence the political decision-making process.

This process resulted in increasing publicity of the debates and discussions, the incorporation of new parties as consultative members and the incorporation of other actors, such as non-governmental organizations, as valid interlocutors. The situation escalated as discussions regarding mining activities in Antarctica advanced and the concern about the potential environmental impacts called the attention from active parties, acceding states and non-parties. In this context, Brazil inserted its policy for Antarctica in a manner that articulated these different aspects, while also defending the maintenance of the Antarctic Treaty (AT) as the basis for the international understanding in Antarctica.

In the previous years, Brazil had clearly delineated its stand in the Antarctic regime, in particular, in relation to the participation with the AT, the need to promote special conditions for developing countries in any agreement on mineral resources exploitation, and the need to give special consideration to environmental protection. For making that compromise more tacit, Brazil needed to finish the process of fully incorporating all the legal framework linked to Antarctica and to update the guidelines for its Antarctic policy, which were more than ten years old and had been made before its involvement in the regime.

The last important step to be included in the complex of the ATS was its accession to the Convention on the Conservation of Antarctic Marine Living Resources (CCAMLR) that resulted from a series of conferences that took place between 1978 and 1980, ending in the agreement signed in Canberra, Australia, on 20 May 1980. Just a little before Brazil's incorporation to the Scientific Committee on Antarctic Research (SCAR), on 8 August 1984, the Presidency presented the proposal to approve the CCAMLR to Congress. In its explanatory memorandum, prepared by the Ministry of Foreign Affairs, Saraiva Guerreiro, it was considered the '…increasing economic interest on Antarctic marine living resources' since that Brazil had met the conditions to access the Commission and had '…concrete interests in the field of Antarctic marine living resources,' stressing the high concentration of krill around the area where the Brazilian station had been established.[4]

[4] Exposição de motivos DNU/DPB/DAI/ SAL/106.10 (D) (1984).

Thanks to its claimed intention to accede to the convention, Brazil was granted permission to attend the meeting that took place in Hobart, in September 1984, as an observer.[5] Present at the meeting was an official of the Brazilian embassy in Canberra and a high official of the Brazilian Superintendency of Fishing Development (SUDEPE). The following year, Brazil participated as observer once more, being represented only by the embassy official. In his speech on the meeting, he declared that Brazil was following closely the work of the Commission and the Scientific Committee of the CCAMLR and that, while it did not have any interest in commercial fishing at that time, it had '…a potential interest in engaging in these activities in the future,' declaring his conviction that the research activities undertaken by Brazil qualified it to become a member of the Commission and the Committee.[6]

In the meantime, the approval of the CCAMLR was being processed within the different commissions involved in the Brazilian Congress. With all assessments favourable, it was approved by the Congress on 5 December 1985, through the Legislative Decree 33/1985. While promulgation still required the issue of a Presidential decree, the decision of the Congress allowed Brazil to request active membership on the convention.

These advancements in the Antarctic involvement were parallel to the increased perception of the importance of the scientific advancement for national development and the necessity of centralizing its administration under the control of the federal government. In March 1985, the Decree N° 91.146/85 created the Ministry of Science and Technology with the intention to overcome the '…disperse and disarticulated…' character of its existent scientific agencies.[7] That resulted in the replacement of the CNPq by the newly created Ministry within the CONANTAR in 1986.[8]

The rapid advancement in organizing a progressive scientific system and in developing a relevant Antarctic scientific programme allowed that, in 1986, Brazil received its first international recognition. In June, during the XIX meeting of the SCAR in San Diego, US, Dr Rocha-Campos was elected Secretary of the SCAR to replace Dr Claude Lorius—who was

[5] Report of the third meeting of the commission (1984, p. 3).

[6] Statement by the Representative of the Government of Brazil (1985, Annex F, p. 3).

[7] Decreto N° 91.146 (1985).

[8] Decreto 92.878 (1986).

on this occasion elected President of the institution—for the remainder of his term, until 1988. While Dr Rocha-Campos was a prominent and well-regarded scientist before the launching of the PROANTAR, his designation as Secretary of the highest international Antarctic science authority was a direct acknowledgement of the expertise developed in Antarctica after just four years of activity and two of being accepted as member of the SCAR.[9]

On 1 August 1986, Brazil presented its request to be included as member of the Commission of the CCAMLR, in accordance with Article VII 2(d) of the convention. By the request of Argentina, a special meeting was convened on the first day of the fifth meeting, on 8 September.[10] The Brazilian delegation included the Ambassador in Canberra, a representative from the Ministry of Foreign Affairs and a scientific advisor designated by the CIRM.[11] After consideration of the credentials of the research presented by Brazil, it was accepted as member of the Commission, being joined by the Republic of Korea and India as new members. The issue of the Decree 93.935/87, on 15 January 1987, which promulgated the CCAMLR, closed the process by ratifying the Convention and inserting it within the domestic legal framework.

Domestically, Brazil was in desperate need of economic resources that could boost its economy. The long record of economic stagnation, inflation and increasing debt coupled with the erosion of the legitimacy of the military regime after two decades of authoritarian rule were increasing social tensions and promoted calls for political reform. The intention to stabilize the economy in the context of high levels of domestic and foreign debt was destined to fail, being unable to control inflation and regain the momentum of production. With its foreign debt growing constantly since 1982, in February 1987, Brazil imposed a moratorium on its debt, forcing the foreign banks to accept losses in capital.[12]

[9] This fact contrasts with the interpretation of Ferreira that acceptance of Brazil as active member of the SCAR was made reluctantly (Ferreira, 2009, p. 139).

[10] Report of the fifth meeting of the commission, 1986. The request for a special meeting was limited to Brazil, while Korea and India were automatically accepted. It is not clear why the Argentine delegation requested a special meeting for Brazil and not for the other two candidates.

[11] They were M.H.C. Cortes, Alcides Prates and Dr. Janice Trotte, respectively.

[12] Dalto (2019).

With Brazil's sights fixed on the economic potential of the region, in May 1987, another round of negotiations on Antarctic mining activities took place in Montevideo. The meeting offered an opportunity for Brazil to partner with China, India and Uruguay, to press the inclusion of special considerations for developing countries, including a guaranteed representation in the different organizations to be included as part of the mining agreement. Therefore, the strategy of avoiding criticism by opening the Antarctic regime to new parties rendered costs to the traditional members that saw an association between the group of the last four consultative members to be incorporated in pushing a new agenda and structuring new relations and distinctions within the Antarctic regime. By proposing a number of amendments equalizing claiming with non-claiming parties and developing with developed countries, the four countries claimed to be oriented to the '...strengthening of the Antarctic Treaty System and the wider acceptance of the Antarctic Treaty as the cornerstone of all activities to be conducted in Antarctica.'[13]

As the consultative meeting to be held in Rio de Janeiro came nearer and the negotiations on Antarctic mineral resources advanced, it was imperative that Brazil updated its National Policy for Antarctica and make it public. On 3 June 1987, the Decree N° 94.401/87 publicly approved the renewed *Política Nacional para Assuntos Antárticos* (Brazil's National Policy for Antarctica—POLANTAR). Whether the General Secretary of the National Security Council considered it merely a formal rewording oriented to make it public, maintaining the involved national security aspects confidential,[14] the new policy presented some important developments from the previous version.

The new POLANTAR acknowledged the centrality that the AT and related agreements adopted for the Brazilian policy for Antarctica, leaving out from the document the national security concerns. The decree even recognized the Convention on the Conservation of Antarctic Seals (CCAS) of 1972 as part of the ATS, when it had still not acceded up until that time—and would only do so between 1990 and 1991. Whether the original POLANTAR stressed the national interests linked with security and economic prospects and recognized the AT as the solely legal valid

[13] Proposed Amendments (1987).
The proposals included the Working Paper 18, 102 and 104.

[14] Atividades da área de política externa (1987, pp. 71–73).

instrument for Antarctica, one of the most relevant changes in the new formula was that it additionally embraced the main objectives of the AT, including the use of Antarctica for peaceful purposes only, the prohibition of any measures of a military nature, freedom of scientific research and cooperation towards that end, the prohibition of nuclear explosions and radioactive waste disposal and the protection of the Antarctic environment.

To the objectives aligned with the principles of the AT, the POLANTAR added the policy objectives of participating within all the components of the ATS, the continuance and the extension of its Antarctic programme, the development of an enhanced knowledge of the region in all its aspects, the development of technological advances applicable to the region and participation in the prospective exploitation of natural living and non-living resources, under special conditions based on its status as a developing country.

Despite all the acknowledgments given to the AT, the document gave Brazil the right to protect '…its direct and substantial interests in Antarctica…' in the eventuality that the Treaty be reviewed. Additionally, the policy sustained that territorial claims should not interfere with the application of the Treaty nor become an obstacle to economic activities, but that the condition of being a developing country should result in special consideration for Brazil. In terms of the implementation of the policy, the decree recognized the role of the CONANTAR and the CIRM within the institutional setting of the PROANTAR, dismissing any mention to any specific institution to be created to that end; this suggests that, by that time, the idea of a Brazilian Antarctic Institute had already been abandoned.

In all, the document expressed the intention of Brazil to keep its options open by committing to the AT and its principles but limiting that compromise to the validity of the treaty. Also, the POLANTAR reflected Brazil's emphasis on natural resources, but with two conditions: the need to consider the conservation of the Antarctic environment, what was important for its possible repercussions over the Brazilian environment and, thus, to its main economic activity; and the need to obtain a special consideration for being a developing country, as opposed to any special consideration derived from a territorial right. In this sense, the economic factor was pivotal for the definition of Brazil's attitudes towards the continent, linking the potential exploitation of the region, the scientific research and the technological developments derived from it and the environmental protection.

Therefore, the relinquishing of any territorial claim and the departure from the tradition inherited from the frontage theory was a calculated move oriented to favour any prospect of economic development in Antarctica. By not restraining its position to a fixed region, Brazil maintained its freedom to access and intervene in any part of Antarctica. At the same time, that option allowed it to insist on replacing any criteria based on territorial rights with the special consideration to the developing world—echoing some of the objections that some non-parties were subjecting the Antarctic regime to. By upholding a dual attitude in front of the ATS—supporting its principles and operation but defending the need of reform and greater openness and reserving undefined rights—Brazil expected to secure the benefits of operating within the system, without compromising any possible benefit derived from its dissolution.

In other considerations, from an institutional aspect, Brazil was signalling a more intense commitment with the scientific aspect of its program. Added to the inclusion of the Ministry of Science and Technology the previous year, in July 1987, theBrazilian Academy of Sciences (*Academia Brasileira de Ciências*) was made part of the CONANTAR, by a representative selected by the Ministry of Foreign Affairs from a shortlist presented by the Academy.[15] Whether the incorporation of such a representative was a significant change in the institutional weight given to each actor or not, it was an acknowledgement of the scientific community independent from the governmental structure.

The increased relevance that the scientific aspect of the programme was gaining was accompanied by the international recognition of its more regarded scientists. In September 1988, Dr Rocha-Campos was re-elected Secretary of the SCAR during the XX meeting in Hobart, Australia, a position he would hold until 1992, evidencing the favourable view of his previous performance and giving Brazil a place of international prominence on the Antarctic science.

It is not clear if the absence of an Antarctic Institute motivated the exclusion of Brazil in what was the first regional instance of the coordination of Latin-American countries in Antarctica, the meeting of the heads

[15] Decreto N° 94.679 (1987c).

of the National Antarctic Institutes of Argentina, Chile and Uruguay held in Buenos Aires in 1987. The inspiration for this meeting came from those of the National Antarctic Programs Directors held in 1986 which were intended to overcome the shortages experienced by the SCAR working group in logistics.[16] The meeting between the three programs would be repeated in 1988 in Santiago de Chile and 1989 in Montevideo, and eventually transformed in the *Reunión de Administradores de Programas Antárticos Latinoamericanos* (Meeting of Administrators of Latin American Antarctic Programmes—RAPAL) with the inclusion of Brazil, Ecuador and Peru in 1990.

Notwithstanding, Brazil did not limit its alliances to the Latin-American region or the Third World. By that time, the Soviet Union was experiencing a crisis resulting from the decline of its regime and economic model, as well as the unintended consequences of the reforms implemented since 1985 by its leader, Mikhail Gorbachev, General Secretary of the Communist Party. The so-called *Perestroika* and *Glasnost* were aimed at dynamizing the economy and a limited democratization of society, which included a greater opening up with the West. This reformulated the Soviet relationship with Brazil, starting a mutual political consultation mechanism in December 1985 and resulting in the first visit of a Soviet Foreign Minister to Brazil between the 28 and 30 September 1987. The visit concluded with a joint communication in which both countries aligned their position on several international policy issues, including a defence of the AT regime and its efficacy in ensuring peace, the denuclearization of Antarctica, the preservation of the environment and the promotion of international cooperation.[17] The following year, a corresponding visit would be paid by Brazilian President Sarney and a similar declaration with an almost identical defence of the Antarctic regime would be issued.[18] These declarations did not result in any special political alignment between Brazil and the Soviets within the Antarctic regime, but made explicit what was considered as the main principles to be maintained: peace, environmental protection—including avoiding any

[16] As described later on, this ended up conforming the Council of Managers of National Antarctic Programs (COMNAP).

[17] Comunicado Conjunto Brasileiro-Soviético (1987).

[18] Comunicado Conjunto Brasileiro-Soviético (1988).

nuclear testing or radioactive waste disposal—and international cooperation, being consistent with the defined defence of the regime established in the recently issued POLANTAR.

With permanent presence in Antarctica, its POLANTAR updated, and already incorporated to the main structure of Antarctic governance, only four years after becoming consultative member, Brazil had the opportunity to host the main meeting of the AT regime. Between 5 and 16 October 1987, Brazil hosted the XIV ATCM in Rio de Janeiro. A Special ATCM to assess the merits of East Germany and Italy to become consultative parties took place on the first day, with a favourable decision for both countries. In contrast to the active role assumed during the mining negotiations, the opportunity of hosting the ATCM did not result in an increased initiative from Brazil. However, the occasion was exploited by the Ministry of Foreign Affairs, Roberto de Abreu Sodré, in his opening speech, to restate the principles and main policy guidelines defined in the recently issued POLANTAR. That included the emphasis on the potential of economic resources, living and non-living, in the region; the reservation of rights in case the treaty was to be revisited; the view that territorial claims should not interfere with economic activities; and the need to consider the situation of developing countries as a special category within any scheme related to mineral exploitation in Antarctica.

Just a few days later, a sudden stock market crash—the so-called Black Monday —affected the world economy, generating a financial crisis of global proportions. The crisis spiralled the inflation in Brazil, aggravating the grave condition of its economy. More than ever, economic resources were wanting, and the potential for Antarctic mining was an opportunity that could be seized, but only if special conditions were granted for countries such as Brazil which, at that time, lacked the technical capabilities to compete with the main powers.

4.2 Mining, Environment and Development

The 1987 Special Consultative Meeting on Antarctic Mineral Resources had arrived at a point where an agreement was in sight, calling for a meeting in the following year in Wellington to sign a convention. Negotiations had progressed since the inclusion of the issue of mineral exploration was pushed into the 1972 ATCM agenda, at that time with strong opposition from Argentina and Chile, who warned that a rush in deciding mineral resources could led to the end of the Treaty.[19] As it turned out, the inclusion of the issue had provoked the involvement of other actors, the interference of non-governmental organizations, and resulted in a significant expansion of the regime, which in that period went from twelve consultative members and other five acceding states to twenty[20] and thirteen states in 1988, respectively.

As a result of that opening of the regime—which also added a number of correlating measures and agreements—the basic distinction of claiming and non-claiming countries was crossed with the distinctions of the communist and Western blocs,[21] and now with the distinction between developed and underdeveloped countries. Thus, the continent for peace and science that had accommodated a single political distinction related to an Antarctic issue—the claimant/non-claimant division—now had to deal with another two divisions that related to distinctions on the international system and that had no link to the actions in Antarctica.

Therefore, a well-known dynamic that had developed for three decades was shaken by the inclusion of new actors, with little—or almost none—previous knowledge of the questions involved, in a moment when one of the most sensitive issues was being discussed. Long years of negotiations and the pressures that the regime was subject to must have exerted a relevant role in fostering the agreement, which now needed to accommodate the positions of the newcomers and avoid outside criticism. Whether the agreement proved, in the short-term, to be destined to fail, it nonetheless reflected the new arrangement and complex dynamics that the Antarctic regime needed to accommodate.

At that point, for Brazil, it was the climax of all its efforts in Antarctica.[22] The country had acceded to the Treaty in view of the potential of mineral exploitation and had invested in its own programme in order to become a consultative party to be able to participate in the decisions

[19] Item 8: Mineral resources (1972); and Antarctic Resources: effects of mineral exploration (1972).

The issue was informally brought about in 1970, but it was only officially dealt with in 1972.

[20] This includes the divided Germany as two different parties.

[21] This division existed previously, but it was quite overshadowed, as originally only the Soviet Union was included in the arrangement.

[22] The rapid development of Brazil's Antarctic presence in Antarctica led Jack Child to consider the country's presence as 'modest but intensifying' (Child, 1988, pp. 133–155).

regarding a mineral regime. That is not to say that Brazil didn't recognize the importance of scientific progress derived from its Antarctic endeavours over long-term development goals,[23] but that its critical economic situation focused its sights on the more immediate interests of natural resources. Whether, as seen, Brazil had also become interested in the prospects of exploiting the living resources under the CCAMLR, it was in the Antarctic mineral resources—especially oil—that Brazil placed its hopes and aspirations. Domestically, in a short time, Brazil had developed its offshore oil exploitation capacities, which by the late 1980s accounted for 70% of its domestic production of that commodity.[24] That significant advancement of the Petrobras offshore capability on Brazil's territorial sea defined a relative advantage in comparison with its regional neighbours and gave rise to expectations about the future potential of the development of the activity in the Antarctic, even when still lacked the technological capability necessary to an enterprise on the Southern Ocean.

During the 5th and 6th Brazilian Antarctic expeditions (1986–1987/1987–1988), the oceanographic vessel *Almirante Câmara*[25] of the Brazilian Navy was involved in geological research on the Antarctic peninsula, surveying the seismic profile of the Bransfield Strait (*Mar de la Flota*) and the Bellingshausen Sea.[26] The expedition was part of an agreement between the Petrobras and the Brazilian Navy aimed to provide the means to the survey of the continental shelf, following the duties assumed as part of the UNCLOS.[27] The agreement, signed in August 1986, determined that the *Almirante Câmara* would be furnished with a multichannel seismic system and other geophysical and navigational equipment to be used in the PROANTAR and the continental shelf survey. The company provided the scientific team, financed the costs of the expeditions, the adaptations of the vessel and the acquisition of the equipment

[23] Fontes Diégues (1987, pp. 105–106).

[24] Decreto N° 92.522 (1986).

[25] The NOc (Oceanographic Vessel) *Almirante Câmara* (previously *USNS Sands*) had a length of 63.45 m, a beam of 12.04 m, displacing 1,448 tons and reaching speeds of 12kt. It had the capacity for 46 people (crew included).

[26] Fontes Diégues (1987, p. 104).

[27] Anexo C, 'Noticiário Marítimo' (2003, pp. 180–189).

from the US Company, Western Geophysical.[28] Additionally, the Brazilian State Oil Company provided assistance at the different stages of negotiations of the Antarctic mining regime through the technical advice of representatives of the company who were present at those meetings.[29]

The survey of the *Almirante Câmara* was another indication that Brazil viewed the Antarctic through the same lens with which it approached its maritime rights—something reinforced by the inclusion of the political question within the *Divisão do Mar, da Antártica e do Espaço* (Seas, Antarctic and Space Division—DMAE)—and that the decision of assigning the responsibility of the PROANTAR to the CIRM had important effects in the way in which the programme was structured.[30] As acknowledged by the CIRM itself, the seismic survey in Antarctica was not only directed to dominate the multi-channel seismic technology, but also to identify the potential of fossil fuels on the Antarctic sea-bed.[31] It is important to acknowledge, however, that Brazil was not the only consultative party undertaking research on the potential of fossil fuels in the Antarctic by conducting geological research. Japan, Norway, US and the Soviet Union had conducted surveys on their own in the 1970s.[32] Also, it was not the only country to take advantage of the counselling from oil companies, as evidenced by the participation of representatives from oil and mining companies in the US and British delegations to the ATCMs.[33] What differed in the case of Brazil was that the oil industry production was under the monopoly of the State Oil Company,[34] which determined a completely different attitude in the mining negotiations.

[28] Besides the crew, there were a team of eleven scientists, eight from Petrobras and three from the Western Geophysical. See: Nunes (2005, p. 130) and 'Noticiário Marítimo' (2003, p. 183).

[29] Ferreira (2009, p. 142).

[30] The emphasis given to natural resources of the Sea within the Brazilian Navy could be seen in: Soares Guimarães (1986, pp. 134–135).

[31] 'Modelo Brasileiro para o Desenvolvimento das Atividades Voltadas para os Recursos do Mar' (1998, p. 52) and Nunes (2005, p. 130).

[32] Villa (2004, pp. 122–123).

[33] Moneta (1986, p. 840).

[34] The oil production and distribution State monopoly in Brazil was established by Law 2004 in 1953. It formally ended in 1997 but continue to be exerted in practice as Petrobras continued to dominate all the chain of production and distribution of the Brazilian gas and oil.

As part of the new dynamic within the ATS, Brazil sought to push its own agenda on the negotiations. It had tried to favour a scheme in which each contracting party would be a partner in mineral exploitation instead of establishing a regulatory scheme for private activities. However, that proposal conflicted with the interests of the most technological advanced parties, such as the US, which wanted to guarantee non-discriminatory access to Antarctic minerals to their corporate nationals.[35] Eventually, the will of the latter prevailed, what would mean unequal access to mineral exploitation to countries such as Brazil due to the technological gap. Brazil also attempted to grant more power to the political body of the Convention, the Commission, instead of the regulatory committees, but it was alone in that effort. Eventually, it had to be content with the recognition of the special position of the development countries and the attribution of a quota of representation on the regulatory committees to developing countries, backed, as seen, by the partnership of the other consultative newcomers, China, India and Uruguay.

After six years of formal negotiations, on 2 June 1988, the Convention on the Regulation of Antarctic Mineral Resources Activities (CRAMRA) was adopted. As many other parties on the occasion, Brazil celebrated the spirit of cooperation and negotiation that allowed it to come to terms in a very sensitive situation. However, the Brazilian head of the delegation, Antonio Guerreiro,[36] manifested in its statement on the occasion of the adoption of the Convention that, as many other parties, Brazil felt that the Convention felt short in many aspects. In particular, Guerrero considered that the agreement was irresolute in taking into consideration the interests of the developing countries and the concerns of humankind as a whole.[37] For Brazil, the agreement fell short of guaranteeing equal access to mineral resources due to the structural differences of development and resulted in another arrangement that favoured the developed nations. Nonetheless, the recognition of the developing countries as a special category and the strict environmental rules applicable were two triumphs that Guerreiro could display as partial successes.

[35] Joyner (1987, p. 894).

[36] Antonio Guerreiro, son of the former Minister of Foreign Affairs Saraiva Guerreiro, was a young diplomat in charge of the Department of International Organizations.

[37] Statement by Mr. Antonio Guerreiro (1988, pp. 26–27).

Designed to be an arrangement that could accommodate many disparate interests, the CRAMRA ended up not satisfying any party. Not internally, not externally. Internally, the claiming countries were left in a position where they would eventually clash—in the case of Argentina, Chile and United Kingdom—or with non-claimant countries on the issue of royalties from the mineral exploitation taking place in their claimed sectors. Besides, the developing countries would be left mainly without the benefits of the activities, due to its technological and capital restrictions—especially considering the deep crisis they were confronting.[38] Externally, the convention set the bar so high that exploitation would be almost impossible in economic terms, while the risk and impacts on the environment would still be a possibility. If we add that no prospects of economic mineral resources exploitation were foreseen in the near future, it is not surprising that the Convention didn't take long to fall.

Despite the relative disappointment with the CRAMRA, when the agreement was opened to signature, on 25 November 1988, Brazil signed the Convention.[39] Nonetheless, doubts started to arise within the Australian government as soon as the convention was finished and, as public opinion became aware of the issue and foreign and domestic environmentalist groups started to call a halt in the agreement, in May 1989 the Australian Prime Minister, Bob Hawke, announced that it would not ratify the agreement.[40] Whether the Australian misgivings in signing the agreement also responded to economic and sovereignty concerns,[41] it was displayed in the public eye as motivated by conservationists' anxieties.[42] The Australian position became a joint one once that, in August, the French Prime Minister, Mr Michel Rocard, announced that his government would also reject the CRAMRA, calling for a comprehensive Antarctic environmental protection scheme. By that time, most of the parties of the Convention had already signed it, but without the

[38] Joyner (1987, p. 894).

[39] The same day the Convention was also signed by Finland, New Zealand, Norway, South Africa, South Korea, Sweden, Union of the Soviet Socialist Republics and Uruguay.

[40] See: Blay and Tsamenyi (1990, pp. 197–200).

[41] As soon as the agreement was concluded, Australia's Treasurer, Paul Keating, urged the cabinet to be careful with the signature of the CRAMRA as it would result on a threat to Australia's claimed sovereignty over 42% of Antarctica and the economic rights derived from it (Morgan, 1988).

[42] See: Blay and Tsamenyi (1990, pp. 199–200).

approval from Australia and France, the agreement was a dead letter.[43] In a joint statement, on 18 August 1989, the Australian and French Prime Ministers declared that it was their intention to promote the protection of the Antarctic environment and its dependent and associated ecosystems, considering that mining activities were not compatible with such protection and that, therefore, they would propose that the continent become a wilderness reserve.[44]

Australia's and France's reaction was also a response to the pressures that the environmental issue was exerting from outside the Antarctic regime. The famous conservationist Jacques-Yves Cousteau had long became involved in a campaign to transform Antarctica in a wilderness reserve, effort that was shared by environmental organizations such as the Antarctic and Southern Ocean Coalition, the International Institute for Environment and Development, the Environmental Defence Fund and the Antarctica Project.[45] Greenpeace even installed, in the Antarctic 1986–1987 season, its own station, named World Park Base,[46] from where it monitored other stations' compliance with the AT environmental regulations. In its report, published in 1988, it stressed that most of the stations had not complied with the provisions agreed by the AT parties in the 1964 Agreed Measures and other regulations, raising doubts about the possibility of the AT parties to impose strict environmental policies to private companies when they were inefficient in imposing them to its own operations.[47]

The situation escalated as, in February 1989, the Argentine vessel *Bahía Paraíso* sank in West Antarctica, spilling a great amount of oil and affecting the surrounding wildlife. The accident was seized on by the Third World countries led by Malaysia in the UN General Assembly to press their demands in front of the AT parties. The next month, an oil transport ship, the *Exxon Valdez*, suffered an accident that resulted in an environmental catastrophe in the vicinity of Alaska . These concerns

[43] Besides Australia and France, the Convention still needed the signature of Japan—which signed it in November 1989—Belgium, India and Italy.

[44] Joint Statement (1989).

[45] Blay and Tsamenyi (1990, p. 198); Joyner (1987, p. 895) and Villa (2004).

[46] It was located on Cape Evans, in the Ross Island, within the so-called New Zealand sector. It was deactivated in 1992.

[47] Villa (2004, pp. 130–135).

led Chile to present, in the preparatory meeting for the XV ATCM to be held in Paris, a proposal to include a comprehensive framework for environmental protection in Antarctica, in order to ensure the environmental responsibility of all human action in the continent and provide with measures for assuming environmental responsibility and reparations in case of accidents and other unintended impacts.

In Brazil, the refusal by Australia and France to sign the mining convention was almost unnoticed in the media, being mostly treated as not involving Brazil.[48] In official circles, the position was to back up the mining agreement. The view was that the convention provided some regulation in case mining activities began at some point, while also avoided the potential political conflict that could arise if mineral exploitation started without a comprehensive agreement, especially when the thirty years period for loosening of the conditions imposed for the revision of the regime was coming to an end.[49]

Internationally, for Brazil, the situation was a delicate one. If, in Antarctica,Greenpeace had acknowledged the environmental performance of the EACF,[50] the country had been subject to intense criticism for its lack of environmental responsibility regarding the Amazon Forest. By the end of 1988, the World Bank was pressing the Brazilian government to adopt measures to protect the Forest, while all the environmentalist groups were hitting the Brazilian lack of active policy in regards the conservation of the nature and the protection of the aboriginal peoples.[51] Whether there were numerous legislations in Brazil regulating a wide range of environmental aspects, its lack of compliance was the result of a combination of a conservative conception of private property, which precluded the State for intervening in private areas; the action of the judicial power that, based on the former liberal ideology, based its decision in an individualistic perspective that precluded the identification of collective

[48] During the research only a single news presented a mention to Brazil's attitude against a total mining ban (Costeau quer bloquear, 1989).

[49] Ferreira (2009, p. 143).

[50] Only three stations received favorable evaluations on the Greenpeace report: Brazil, Italy and Poland (Villa, 2004, p. 131).

[51] Bird Denuncia destruição (1988); and Bird suspendió recursos (1988).

damage—such as that produced by environmental damage—and, fundamentally, a lack of an environmental policy, understood as a coherent and integrated action at all governmental levels.[52]

Nonetheless, by October 1988, the country approved a new democratic constitution that ended twenty years of dictatorship. One relevant change introduced by the new constitution was the inclusion of the environmental question, probably strongly influenced by international pressure and the action of environmentalist groups in the domestic setting. The new basis of the democratic regime included a complete chapter dedicated to the environmental protection. Making the Brazilian State and the community responsible for the safeguard of the environment, Chapter VI of Title VIII of the 1988 Brazilian Constitution recognizes the society's right to a balanced environment, considered as a common good, basis for the health and quality of life of the general population. In pursuing that objective, it granted the State broad competences for the evaluation, control, the attribution of responsibilities and the protection of the general ecological balance and the preservation of regions of particular environmental value.

This renewed attitude of Brazil regarding the environment was compatible with the protectionist attitude of Australia and France in the Antarctic, but it confronted with its short- and medium-term needs for economic development. To make the issue more complex, whilst Brazil was interested in the possibilities that oil exploration could bring it in the future, its more immediate priority was not altering the ecological and climatological balance between Antarctica and the country's weather, on which its main economic activity depended—and which the new constitution made the State guarantor of. For Brazil, the CRAMRA was acceptable precisely because it combined the possibilities of obtaining some gains from the economic activity while also ensuring the protection of the environment; a perspective that was shared with the other Latin-American countries. However, with the lack of a regulatory agreement—something that the attitude of Australia and France had guaranteed—the negotiation of a wide environmental framework that could ensure the ecological balance of Antarctica was the second-best acceptable option, compatible with the recently acquired environmental responsibility of the Brazilian State.

[52] Fernandes (1992).

Different from the British and the US, where important corporate interests lobbied for the maintenance of the CRAMRA, the Brazilian governmental monopoly on the activity facilitated that the CRAMRA be more rapidly discarded. Thus, when the other consultative parties met in Paris at the XV ATCM, Henrique Valle, head of the Brazilian delegation, declared that, as other parties had decided to withdraw from the agreement, Brazil would not insist on it, even when it disagreed with the possibility that any activity related to mineral resources could be conducted without a regulatory scheme. Added to the potential conflicts that such activities could arouse, the Brazilian representative highlighted the importance that Brazil gave to the protection of the Antarctic environment due to distance between that region and Brazil, stating that such preoccupation with the Antarctic environment was not new, nor aroused from the influence of the international opinion.[53]

The Paris XV ATCM agenda was almost completely concerned with the environmental issue, as the time for the AT to provide a response to the international community was closing. A UN resolution obtained by Malaysia and its allies from the Non-Aligned Movement insisted on the need to include the whole international community in discussions regarding Antarctic mineral resources and lamented that the AT parties advanced with the convention despite the call for a moratorium by the UN.[54] That call acquired more relevance before the break-up of the consensus of the 1988 Convention and the increased notoriety of the environmentalist organizations claim in the light of the *Bahia Paraíso* and *Exxon Valdez* accidents. Furthermore, the UN decided to convene a new Conference on Environment and Development no later than 1992,[55] which aroused fears that the Conference could try to interfere with the Antarctic if the AT parties failed to provide some credible environmental framework. With that in mind and being unable to reach any consensus on the matter, the ATCM decided to call a Special Consultative Meeting to be held the following year in Viña del Mar, Chile, to debate a comprehensive environmental framework for all human activity in Antarctica. However, and to leave all options open, the parties also agreed to hold

[53] Opening Address (1989).

[54] Resolution 43/83 (1988).

[55] Resolution 43/196 (1988). See also: Bemejo García and Bou Franch (1993, p. 53).

another Special Consultative Meeting in 1990 to discuss the liability protocol required by the CRAMRA, that would not eventuate.[56]

Those developments were taking place in a world system that was changing dramatically. Since 1988, a series of crises within the Soviet Union resulted in the so-called 1989 Revolutions in Poland, Hungary, East Germany, Bulgaria, Czechoslovakia and Romania; concluding with the dissolution of Eastern Europe, the fall of the Berlin Wall in November 1989, and the reunification of the country the following year. The alteration of the order that characterized the world for more than four decades introduced additional uncertainties within which the Antarctic regime should navigate in order to stay relevant. Progressively, the idea of a global order on a planetary scale arose from the environmental consciousness and the imminent end of the division between the Eastern and Western blocs.

Those changes also impacted in Latin-America, which was advancing in a process of regionalization and abandoning the competitive attitudes of the 1980s.[57] As seen, in June 1990, Brazil, Ecuador and Peru joined Argentina, Chile and Uruguay and formed the RAPAL.[58] Being the first regional coordination setting, the RAPAL was a by-product of the creation of the Council of Managers of National Antarctic Programmes (COMNAP) in September 1988. The COMNAP mission was to increase the safety and efficiency of Antarctic operations and resulted from the Meetings of National Antarctic Programmes Administrators between 1986 and 1987. Those meetings were a response to the increasing number of actors in Antarctica—most of them inexperienced—and the limitations of the SCAR Working Group in Logistics when it came to make its decisions and recommendations applicable in field operations. The main limitation was that, being a non-governmental scientific organization, the SCAR lacked any authoritative capacity to make its recommendations mandatory and foster its implementation by the respective Antarctic programmes.[59] As the efficiency and safety of operations

56 Bemejo García and Bou Franch (1993, p. 163).

57 See: Colacrai (2003, 2012, pp. 88–91).

58 Colombia participated as an observer as it didn't have an Antarctic programme at this time. It acceded the Antarctic Treaty in 1989 and it has been present on expeditions of other countries until 2014 when it undertook its first Antarctic Expedition. Up until now (2021), it has not become a Consultative Party of the Antarctic Treaty.

59 Fowler (2000, Chapter 2).

were closely linked to the environmental question, from its inception the COMNAP included the need to equalize the environmental performance and application of rules and procedures within its objectives.[60] That same approach was followed by the RAPAL which in its first meeting issued a declaration in which the participants stated that:

> The protection and conservation of the Antarctic environment and its dependent and associate ecosystems is an issue of global relevance, especially for the neighbouring continent countries, within which the Latin-American countries present in this meeting could be accounted.
>
> That, for such a reason, it is necessary to support and promote any measures aimed to such protection and conservation, articulating, in a due and harmonic manner, the legitimate uses of the Antarctic area within the framework of the Antarctic Treaty System, with special emphasis in the use of the region for peaceful purposes only, including the denuclearisation, freedom of scientific investigation and the interchange of information in a system of international cooperation.[61]

Formal and informal meetings between the AT parties mediated between the XV ATCM and the XI Special ATCM, resulting in four articulated proposals on the Protection of the Antarctic Environment that were presented as basis for discussions. Added to the establishment of the RAPAL, in joint declarations with Argentina and Chile, Brazil assumed the commitment to coordinate its position with those countries at the upcoming Special Consultative Meetings to negotiate a comprehensive framework for environmental protection in order to ensure the preservation of the continent.[62]

The XI Special Meeting took place in Viña del Mar, Chile, between 19 November and 6 December 1990, during which the proposals for the environmental framework rose to six.[63] Brazil did not author any of those proposals and it is unclear what its stance was on the different discussions. In his inaugural speech, Mr. Henrique Valle was limited to state

[60] *Ibid.*, Chapter 4.

This would eventually be reinforced by the highlighting of the differences on environmental performance and compliance with regulation raised in the *Greenpeace* report.

[61] Declaración Final (1990, own translation).

[62] Comunicado Conjunto (1988); and Declaração Conjunta (1991).

[63] Bemejo García and Bou Franch (1993, pp. 181–182).

the '...priority we [Brazil] assign to the question of the environmental protection as a whole, and in particular in the Antarctic Continent' and to define that, whether Brazil had its own preferences, it came to the meeting with a flexible attitude in order to attain concrete outcomes, in the hope that they could become part of a positive agenda in the next UN Conference on Environment and Development that the country was to host in 1992.[64]

In respect of the formal aspects, the proposals diverged on if a separate agreement/convention—as it had been the common practice within the ATS—or the adoption of some instrument that would be considered as constitutive of the AT—such as a protocol—would be adopted. In the material aspects, besides codifying and unifying the already existing environmental regulation, it was pending the adoption of a system of arbitration and liabilities and the question of whether to maintain the CRAMRA or adopt some form of prohibition of commercial mining activities—and, in that case, which form and extension. The Viña del Mar meeting did not reach a consensus on any of these issues but produced important advances and resulted in a first draft of Protocol, elaborated by Mr. Rolf Trolle Andersen, from Norway, that condensed those advancements.[65]

By that time, the Brazilian Antarctic programme was experiencing important setbacks. The first Constitutional President after the military dictatorship, Fernando Collor de Mello, inherited a deep economic crisis, characterized by a fiscal imbalance, debt crisis and inflation which had skyrocketed. As many of the other Latin-American governments, Collor de Mello started a programme of reforms inspired by the so-called Washington Consensus[66] that included State spending cuts, the privatization of State companies and exchange rate liberalization, among others. That meant the development project based on heavy State investment and intervention was abandoned and a neoliberal *ethos* was adopted.

That impacted significantly on the State-sponsored programmes with no immediate economic benefits, such as the PROANTAR. The *Besnard*

[64] Opening statement (1990).

[65] Bemejo García and Bou Franch (1993, pp. 188–190).

[66] Here, the concept is used in its usual common meaning that refers to a set of a neoliberal reforms. A discussion of how the concept arose and was appropriated to reflect a set of policies that were not originally included is present in Williamson (2009), but has no practical effects on our analysis.

ended its Antarctic expeditions in January 1988, after a fault in the propeller shaft left it adrift in the Drake Passage. In June 1989, a Brazilian company, Engenavi, secured a contract for the construction of a sophisticated polar vessel—a project that had been developed since 1985, but which the economic crisis had prevented from starting construction. In 1990, the programme also had to deactivate the shelter Padre Balduíno Rambo, due to a problem of maintenance and diplomatic reasons.[67] The spending cuts also led the programme to abandon any plans for establishing a second station and to suspend the works for the extension of the EACF as well as the inclusion of new research projects, with the exception of the instalment of an automatic meteorological module some 1,000 m from the EACF.[68] Following the new *ethos*, the Collor de Mello government even entertained the idea of involving Universities and private companies in financing the programme.

This change of perspective in the role of the State also influenced Brazil's foreign policy that departed from the previous principles linked to the predominant developmentalist paradigm. In Antarctica, that resulted in the change of the Brazilian position regarding mining, as it started to explicitly reject any commercial mining activity and support a permanent ban, limited for the possibility of revision under special circumstances.[69] That change in attitude before the second round of the XI Special ATCM was celebrated by Greenpeace, which insisted on the need of gathering a strong coalition favouring a permanent ban on Antarctic mining.[70] Additionally, to offset the cutting of investment and give publicity to the programme, President Collor de Mello and his wife visited the EACF on 20 February 1991, accompanied by the Minister of the Navy and a large entourage, the first visit of a Brazilian President to Antarctica.[71]

Probably, as a signal of its commitment to the conservation on the Antarctic environment, in March 1991, Brazil ratified the Convention on the Conservation of Antarctic Seals (CCAS) of 1972. The CCAS was

[67] Allegedly, there were incidents with the Chilean Marsh station, but during the research no evidence was found of such conflicts nor the solely source of that information provides any details (See: Alvarez, 1995, p. 124).

[68] Proantar precisa de verbas (1991).

[69] *Ibid.*

[70] McTaggart (1991).

[71] *Brasil na Antártica: 25 anos de história* (2008, pp. 65–67).

acknowledged by the 1987 POLANTAR, but only in October 1990 did the Brazilian Congress ratified the Convention that was approved by the President on 18 March 1991.[72] Whether there is no clear indication of the motives to adhere to the Convention at that time, it is probable that while the CCAMLR was accessed with more urgency in view of the potential of fishing in the region—specially krill—the absence of any prospects for sealing activities precluded any hurry in the case of the CCAS. However, in the light of the relevance that the conservation of Antarctic environment acquired internationally in the turn of the decade, the adherence of all consultative parties to the mechanisms of environmental conservation acquired more relevance.[73]

Continuing with the efforts to provide some form of comprehensive environmental framework for Antarctica, the consultative parties met again, this time in Madrid, Spain. Whether the drafted protocol of the previous session was intended to be a basis for discussion, during the second round of the XI Special ATCM in April 1991 it ended up becoming a working draft Protocol, settling the formal aspect and the core content of the agreement. Over that basis, the most contentious issues at the meeting were related to the commercial mining activities, including if an undefined ban or a limited prohibition was to be established, and what mechanism would be used for modification or amendment of the agreement. At the end, the option chosen was an undefined prohibition and a very restrictive mechanism to introduce modifications or amendments, which provoked a last-minute US refusal to the agreement.[74] That led to a third round at the beginning of October where a conciliatory formula was agreed. The new agreement included an undefined ban on commercial mineral activities and a fifty-year period in which the Protocol could not be modified, after which any party could call a revision convention which modifications could be approved by a more flexible formula.[75] The US block of the agreement made the Protocol

[72] Decreto Legislativo 37 (1990); and Decreto 66 (1991a), respectively.

[73] Ferreira assesses that the Brazilian government considered the CCAS as a dead letter (2009, p. 140). However, there would be no reason to access and ratify a convention if it was considered out of rule. The difference between the Brazilian attitude in relation to the CCAMLR and the CCAS has, probably, more to do with the potential economic interest of exploitation of each resource than any juridical consideration of its value.

[74] Bermejo García and Bou Franch (1993, pp. 199–201).

[75] A detailed analysis could be found in Bermejo García and Bou Franch (1993).

felt short from being agreed before the 30th anniversary of the enter into force of the AT, but just in time before the XVI ATCM in Bonn. On 4 October 1991, the Protocol of Environmental Protection to the Antarctic Treaty (also known as Madrid Protocol) opened for signature, inaugurating a whole new phase in the Antarctic governance.

4.3 Under the Steering of Science

By the end of 1991, three forces converged to change the direction of Brazil's Antarctic policy: alterations in the world order, domestic changes with the consequent reorientation of Brazil's foreign policy and the important changes that the Protocol of Environmental Protection introduced in the Antarctic regime. The substantial changes within the international system, with the dissolution of the Soviet bloc and the end of the Cold War and the logic of bipolar competition, altered any strategic significance that the Antarctic region may have had in the past. If Brazil officially never subscribed to the view of the frontage theory on the military strategic relevance of the region and had abandoned any mention of it when elaborated the 1987 POLANTAR, the end of the Cold War made any reference to it inconsequential. Furthermore, the new international situation coincided with the domestic return to democracy, the so-called promotion of the Washington consensus policy in Latin-America, and the Brazilian effort to regain the international credibility by subscribing to the international rules established by the world's centre of power, setting aside the developmentalist policy and discourse and subscribing to the environmentalist concerns that arose globally. Lastly, these events were coupled with the significant change in the priorities that the Protocol of Environmental Protection imprinted in the Antarctic international regime, making the environment concerns top priority.

At the same time, these changes were taking place within the context of increasing regional integration, as the framework of the *Mercado Común del Sur* (South American Common Market—MERCOSUR) advanced. This explains why the positions and alignments that had Brazil shown within the Antarctic regime now changed significantly. The 1991 ATCM in Bonn reflected the recent agreement in RAPAL with proposals submitted by the group of six Latin-American consultative parties members of the RAPAL. Additionally, the previous references to the status of Brazil as a developing country disappeared, as was shown by the absence of any reference in the opening address of the Brazilian

Head of Delegation, Ambassador Enrique R. Valle, and subsequently.[76] This was replaced with a recognition of the efficacy of the Antarctic regime in coping with increasing participation and worldwide attention and providing a wide environmental framework which was considered an arrangement that '...alter[s] in a fundamental way the functioning of the Antarctic Treaty System, without tampering with its fundamental premises and objectives.'[77]

The absolute priority given to Antarctic environmental preservation and the consequent recognition of the Protocol of Environmental Protection was presented as part of a commitment of Brazil to the environment, in conjunction with the fact of being the host of the upcoming United Nations Conference on Environment and Development. These attitudes were evidently directed to raise the international profile of Brazil in the environmental question after receiving much criticism in the previous years on its management of the Amazon Forest. But, while the Amazon would remain a source of natural resources and a contentious arena between conservationists and economic interests, the Antarctic conservationist attitude would continue to inform the Brazilian policy for the white continent, becoming one of its most characteristic traits. This was also reflected in the incorporation of the recently created Secretary of Environment into the CONANTAR in May 1991.[78]

However, the most relevant change in Brazilian Antarctic politics was the transference, in 1991, of the scientific responsibility of the PROANTAR from the CIRM to the CNPq.[79] Up until then, the CIRM issued the invitations to present research proposals to the programme and, even when the CNPq provided its expertise in the evaluation of the scientific aspects of the proposals, it was the CIRM Secretariat who had the authority to select the projects to be implemented.[80] This change did

[76] A detailed content analysis of the Brazilian opening discourses in the ATCMs can be found in Cardone (2015).

[77] Opening address (1990).

[78] Decreto N° 123 (1991b).

The Secretary, depending directly on the Presidency, was created in 1990. Previously, it depended on the Ministry of Interior (since 1973) and, later, was integrated as the Ministry of Urban Development and Environment (1985).

[79] Câmara et al. (2021, p. 3); Gandra (2013, p. 106).

[80] It must be remembered that when established the National Committee on Antarctic Research, in 1984, the CIRM Secretary was named coordinator.

not alter the centrality that the Navy, through the CIRM, had within the PROANTAR, nor the control it exerted in selecting the projects under the logistical aspects of operations. However, it was an explicit recognition of the specifics of the scientific activity and the competences of the scientific national agency in what was regarded as the evaluation of the scientific aspects of the proposals. Furthermore, it signalled the abandonment of any intention of using Antarctic scientific research for 'strategic' purposes—especially those related to the potential use of natural resources of the region—and the attribution of complete autonomy to the research to be developed in Antarctica, at least in regard to the definition of its objectives and areas of application.

The transference of the responsibility included the need to fund scientific activities through the CNPq, while the cost of logistics remained to be funded by the Navy through the CIRM.[81] This change in the financial aspect produced important conflicts in the funding of scientific activities, even threatening to halt the work developed until then and motivating the withdrawal of one project, which had been running for ten years.[82] Such instability of Antarctic scientific funding became one of the more prevalent traits of the PROANTAR.

The Antarctic programme continued to receive public support by the visit of the Vice-President Itamar Franco in January 1992.[83] Notwithstanding, domestically, the political landscape was one of increasing unease. By the middle of 1992, a Parliamentary Commission was instituted with the objective of studying accusations against the President Collor de Mello. The irregularities found during the investigation motivated a wave of popular protest and, on 1 October, an impeachment process was brought to the Senate. The process resulted in the renouncement of Collor de Mello to the Presidency on 29 December, notwithstanding, the following day, the Senate decided for the destitution of the former President and determined its ineligibility for eight years. The elected Vice-President and interim President since the start of the impeachment, Itamar Franco, assumed the Presidency.

[81] Additionally, Petrobras continued to provide the fuel for all operations at no cost.

[82] Expedição à Antártida (1991); and Brasil está fora (1992).

[83] In fact, one of the complaints of the scientific community was that the programme was increasingly used for visiting the region instead for research. See: Brasil está fora (1992).

The same day as the resignation of Collor the Mello, another milestone was reached by the PROANTAR. A Brazilian Air Force Hercules C-130 made a parachute cargo airdrop in the vicinity of the EACF, inaugurating a practice that would assist with the provision of the station during winter ever since.[84] Adding to the possibilities that the new capacities brought to the winter operations for the Brazilian programme in Antarctica, the practice resulted in the acquisition of a new capacity for the Air Force that could be used in different situations if needed.

As regards Brazil's position within the Antarctic regime, following the signature of the Protocol of Environmental Protection, it centred its agenda at the ATCMs in Venice, Italy (1992) and Kyoto, Japan (1994), in three main points: the agreement on what was seen as a key component for the implementation of the Protocol, the Annex on Responsibility; the installation of an Antarctic Treaty Secretariat, deemed necessary due to the increased volume of information and work needed for the operation of the regime; and a discussion on tourist activities. All three aspects were important to the Protocol and were considered by Brazil as essential to make the agreement effective.

Despite Brazil continuing to struggle with important economic challenges, including skyrocketing inflation, by 1994, the PROANTAR started to show some signs of recovery. In the long run, after a couple of years of serious hardships, the transference of the scientific responsibility to the CNPq resulted in a growth of funding for scientific activities, an increased number of projects and researchers involved, and a significant improvement of scientific production since 1993–1994.[85] That is not to say that the funding issues disappeared, as scientists continued to experience limitations of funding and uncertainty, limitations that were not felt at the same level by the Navy in their part of the programme.[86]

Notwithstanding the economic hardships, the *Barão de Teffé* had already reached the end of its operative life to operate in polar waters safely, and the Navy opened an international call for proposals for its replacement in December 1993. In February 1994, the Norwegian vessel *Polar Queen* was assessed by officials of the Brazilian Navy in South Africa,

[84] Nunes (2005, p. 136) dates the event to 28 December, but numerous sources place the event on the 29th. See for example: *Brasil na Antártica: 25 anos de história* (2008, p. 71).

[85] Avaliação preliminar do Programa Antártico Brasileiro (2006, pp. 27–35).

[86] Brasil completa 12 anos (1995).

resulting in its acquisition for USD 15.9 millions.[87] The *Barão de Teffé* had its last Antarctic voyage during the 1993–1994 season, closing its participation as part of the PROANTAR on 2 April 1994 after twelve seasons in the Antarctic.[88] It was replaced by the *Polar Queen*, incorporated into the Brazilian Navy under the name *Ary Rongel*,[89] known by the Navy personnel as the *gigante vermelho* (red giant).

The new vessel had a displacement of 3,628 tons and a length of 75.2 m and was able to transport up to two helicopters and a total of 105 people, including seventy-eight crew members.[90] The ship represented a mayor update in technology but did not result in the acquisition of new capabilities. However, it was a necessary step to maintain the Brazilian operations in Antarctica within the necessary conditions of safety, especially considering the responsibilities assumed to possible environmental impact as part of the Protocol of Environmental Protection and the experience of the Argentine vessel *Bahía Paraíso*. In addition to the new vessel, the PROANTAR developed a heliport in the vicinity of the station to facilitate the logistic activities.

Adding to the importance of the scientific aspect of the Brazilian Antarctic programme, the geologist and scientific coordinator of the programme, Professor Rocha-Campos was elected president of the SCAR, a position he held between 1994 and 1998, becoming the only Latin-American citizen to do so. Whether the instability experienced in the funding of scientific activities rendered the programme vulnerable to the evolution of the economic and political landscape, the notoriety acquired by the programme scientific references and the international impact of its work ensured some prominence to the PROANTAR.

[87] Nunes (2005, pp. 136–138) and 'Noticiário Marítimo: NApOc Ary Rongel' (1994). Nunes locates the supervision of the ship in February 1993 instead of 1994, however, the version of the Revista Marítima Brasileira seems more likely.

[88] It would continue to serve in the Brazilian Navy in signalling tasks of the Brazilian coast.

[89] The NApOc (Oceanographic Support Vessel) *Ary Rongel* have a length of 75.2 m, a beam of 13 m, displacing 3,628 tons and reaching speeds of 14.5kt. It had the capacity for 105 people (crew included).

[90] Information obtained from the official Brazilian Navy Website. Source: https://www.marinha.mil.br/secirm/proantar/navios. Accessed on: 10 December 2020.

As Brazil was progressively controlling its economic difficulties through the new economic programme, the *Plano Real*,[91] implemented by Itamar Franco's Ministry of Economy, Fernando Henrique Cardoso, the regional framework of MERCOSUR advanced. The interest in the region also reflected a more active participation of Brazil within the regional Antarctic coordination setting, the RAPAL. After its initial involvement in 1990, Brazil had mostly neglected the RAPAL, being absent from the II Meeting in Montevideo in 1991[92] and the III Meeting in Quito in 1992,[93] and having no substantial participation in the IV Meeting in Lima, in 1993. It is likely that such absence could have had much to do with the financial difficulties and the changes of the institutional organization of the programme, coinciding with the years that, after the transference of the scientific responsibility from the CIRM to the CNPq, the PROANTAR experienced its first serious financial problems. In any case, in the IV meeting in Lima, Brazil offered to host the next meeting, which took place in September 1994 in Brasília. One of the most significative contributions to the meeting was the submission of the Brazilian-Polish proposal to designate Admiralty Bay as an Antarctic Specially Managed Area (ASMA), following the designation of special areas defined by the Protocol of Environmental Protection, and which eventually became the first of such areas.[94]

On 1 January 1995, Fernando Henrique Cardoso, who had been considered responsible for the economic success of the *Plano Real*, acceded to the Presidency. Under Cardoso, the PROANTAR continued along the same basic lines, particularly in relation to the adoption of the Protocol for Environmental Protection and the measures taken to assume the responsibilities derived from the agreement. Thus, to adapt its institutional design to the new demands, in January 1995, the *Grupo*

[91] The *Plano Real* was an economic programme oriented to stabilize the economic variables, including the devaluation of the currency and the inflation. It included some policies based on neoliberal reforms, other heterodox measures, and the introduction of a new currency, the *Real*.

[92] Declaración de Montevideo (1991).

[93] That perception derives from the report of the meeting available at the RAPAL website. However, it is important to highlight that the report was not the official proceedings of the meeting. In any case, there is no single mention in the document of contributions from Brazil, indicating that, at best, its participation was negligible. (Source: http://www.rapal.org.ar/RAPAL3.DOC. Accessed on: 10 December 2020).

[94] Antarctic Specially Managed (1994).

de Availação Ambiental (Environmental Assessment Group), under the coordination of the Ministry of Environment, was incorporated into the structure of the PROANTAR within the CIRM. The newly incorporated sub-commission was added to the existent Operational Group (*Grupo de operações*), coordinated by the CIRM, and the Advisory Group (*Grupo de Assessoramento*), coordinated by the CNPq since 1991. Additionally, on 6 June 1995, Congress approved the text of the Protocol of Environmental Protection by the Legislative Decree N° 88, following the favourable judgement given in 1993 by who was at that time the Minister of Foreign Affairs, Fernando Henrique Cardoso.[95]

Despite the Protocol not having entered into force, Brazil, as the other consultative parties, was committed to its fulfilment. In the season 1995–1996, the Brazilian Antarctic programme, after consultations with the British government, proceeded to dismantle and complete the removal of the abandoned British station G in the vicinity of the EACF, trying to avoid any potential environmental impact from its deterioration. Also, it advanced the project for the establishment of the ASMA Admiralty Bay, presented in 1996 jointly with Poland with the support of Ecuador and Peru.

In other considerations, continuing with the intention of granting an increasing scientific facet to the programme, the creation of the *Comitê Nacional de Pesquisas Antárticas* (National Committee on Antarctic Research—CONAPA), in 1996, underpinned the scientific character of the Brazilian Antarctic Policy.[96] The newly created committee was assigned to advise the Minister of Science and Technology, the CONANTAR and other government bodies involved in the Antarctic programme on matters of scientific and technological interest and representing the country in front of the SCAR.

The institution of the CONAPA signalled the relevance that the scientific aspect acquired, expressed in its conformation as well as in its place in the institutional setting. Its composition was mainly to be scientists and representatives from the scientific community, including two representatives from the Ministry of Science and Technology, two from the CNPq and up to six scientists with a recognized track record of Antarctic research. Additionally, and adopting the inter-ministerial character of all

[95] Mensagem N° 231 de 1993 (1993).

[96] Decreto N° 1.791 (1996).

the bodies linked to the Antarctic programme, the Committee would be formed of two representatives from the CIRM Secretariat, one from the Ministry of Foreign Affairs and the coordinator of the Environmental Assessment Group. The coordination and vice-coordination of the Committee would be in charge of the representatives from the Ministry of Science and Technology and the CNPq, respectively.

More significant is the fact that the CONAPA was established outside the umbrella of the CIRM, notwithstanding most of its members were part of the different groups that formed the subcommittee of the PROANTAR within its Secretariat. At least in form, the CONAPA was parallel to the CONANTAR in advising the political decision-making and the CIRM in aspects of scientific expertise. This was a significant advancement from the original establishment of the Committee as part of the CNPq, however, under the coordination of the CIRM Secretary and with no official place in the institutional structure of the Brazilian policy for Antarctica—it must be remembered that the Committee was created with the sole purpose of representing Brazil before the SCAR in order to apply for membership. Also, it was a relevant step forward from the attribution of the scientific responsibility for the programme to the CNPq without a meaningful involvement of scientists and scientific institutions in the definition of the Brazilian policy for Antarctica.

The fact acquired even more relevance when taking into account that since 1995 the CONANTAR would not meet, being inactive for ten years,[97] making the CONAPA the only active advisory committee at its establishment. In addition, the view of Antarctic research having strategic importance could be identified in the inclusion in the National Defence Policy of 1996, albeit in a very modest way, the objective being to: '...promote the scientific knowledge of the Antarctic region and the active participation on the process of decision-making of its future.'[98]

However, only two years after its creation, in 1998, the same year that Cardoso ratified the Protocol on Environmental Protection and it came into force, the CONAPA would be deactivated, remaining inactive until

[97] Ferreira (2009, p. 149) and Ata da 161 Sessão (2006, lines 907–908).

[98] Política de Defesa Nacional (1996), Subitem s—Item 5, own translation. Also: Mattos (2014, p. 178).

2004.[99] In that way, the Brazilian policy for Antarctica was mainly channelled through the CIRM Secretariat, leaving the political representation in the international forums to the Ministry of Foreign Affairs and the scientific evaluation of proposals in hands of the Ministry of Science and Technology through the CNPq. This possibility of the Brazilian Antarctic programme to be supported by the CIRM every time its political importance came into question would also be a prevailing characteristic that would project until today, enabling its continuity, but at the same time limiting its evolution as a scientifically driven project.

Ferreirapoint out that the deliberations of the CONANTAR were made through the CIRM, which coincided in great extent in its internal composition.[100] However, the same could not be said of the CONAPA, and its deactivation after only two years of activity reveals the limitations that the political importance of the scientific aspect of the programme was subject to. Furthermore, the deactivation of both committees was an indication of some neglect over the Antarctic programme, at least in its institutional and political aspects. This could be attributed to the convergence of different factors, including the enforcement of the Madrid Protocol, putting an end to the basis of interest in economic resources for a minimum of fifty years; the relaxation of the international pressures on Brazil, relative to environmental issues; the economic strains suffered after the economic international crisis of 1997 and 1998; and the reach of a limit to a first generation of researchers and projects and the need for renewal.

On the other hand, in those years, Brazil's participation as part of the RAPAL intensified, both in terms of the size and specialization of its delegations, as well as the submission of documents and proposals. In 1999, Brazil hosted the X RAPAL in Brasília. In the opening speech, the Brazilian Navy Commander, Sergio Gitirana Florencio Chagasteles, praised the collaboration and advancements made through the RAPAL and its effects on the other institutional instances of the ATS, and the efforts undertaken by all Latin-American countries to protect the Antarctic environment and comply with the provisions of the Protocol of Environmental Protection.

[99] Comitê Nacional de Pesquisas (2004).

[100] Ferreira (2009, pp. 149–150).

The end of the century and the turn of the millennia would not be propitious for the PROANTAR. The economic difficulties started after the so-called crisis of the 'Asiatic Tigers' in 1997 and, after a brief recovery in 1998 following Cardoso's re-election, the crisis in Russia and the following international crisis deteriorated the Brazilian economy, forcing Cardoso to look for the support of the International Monetary Fund under the demand of deeper spending cuts and administrative reforms. The spending cuts and the difficulties experienced by the private sector reverberated strongly through the scientific sector, which increasingly depended upon private contributions. It would take until the end of Cardoso's second term, with the alleviation of the economic difficulties, and the more active involvement of the Ministry of Environment, for the Brazilian Antarctic programme to renew, starting a process of institutionalization and structuration that would consolidate the scientific profile, even when many of the previous traits prevailed.

4.4 Conclusions

Brazil's ambiguity regarding its interests in Antarctica guided its political attitudes about the continent, expressing a dual character in several aspects. First, it oriented its political definitions towards a defence of the AT regime and its principles, however, limited to special conditions, and always leaving the door open to the possibility of review or dissolution of the agreement. In this regard, the 1987 POLANTAR is revealing, as it expressed Brazil's commitment to the high principles of the AT but limited to its rule period. This reservation in the POLANTAR also expressed the inherited influence of the geopolitical thinking and the territorial imprint of the frontage theory, although in a very subtle manner.

Also, Brazil's position regarding the environmental protection of the Antarctic environment was shown to be of a very special nature. The study of Brazil's position during the second half of the 1980s suggests that Brazil's high praise of the need to protect the Antarctic was almost always subordinated to its economic interests. Whether Brazil foresaw the possibility of participating in mineral activities in Antarctica, it defended the need to provide special measures of protection and conservation, notwithstanding that was justified by the necessity to avoid possible detrimental environmental effects on the Brazilian mainland economy.

This emphasis on the economy not only shaped Brazil's position towards the protection of the Antarctic environment, but also its attitude towards the regime. Without a specific interest in any of the Antarctic economic activities but looking to guarantee any potential future benefit, Brazil pursued an open-ended policy that could allow it to be involved in the decisions and remain informed of the advancements and developments of each issue. Thus, accession to the CCAMLR was not necessarily motivated by a desire to be included in all agreements of the regime—as the late accession to the CCAS reveals—but to be included in a scheme that could be of value to the country in the near future. Also, its position in favour of a mining agreement also suggested possible participation in future, even though the country had no technology or expertise in polar regions and would need special conditions for it to be of benefit.

That ambiguity made possible that, once the CRAMRA fell, the more immediate interest in protecting the national economic primary activity from harmful interference of environmental impacts in Antarctica prevailed, making it possible for Brazil to accept a long-term moratorium over mineral activities.

Within the regime, Brazil also pursued a pragmatic scheme of alignments, first partnering with the other newcomers, China, India and Uruguay in pursuing a new distinction between developed and developing countries—and that could overlap the pre-existent distinction between claimant and non-claimant countries. This alignment was replaced with a more common coordination with the other Latin-American Antarctic countries, once the Protocol on Environmental Protection was in place and the RAPAL was established.

At last, the internal evolution of the Antarctic policy of Brazil also reflected the contradictions of its scientific policy and the shift in the political importance of the programme. The formal and political relevance given to the issue was not always accompanied by the respective financial resources, subjecting the activity to important uncertainties. Furthermore, notwithstanding the declared centrality of scientific activity and the institutional advances which gave a hierarch to the scientific elements of the programme, the difference with which the scientific and logistical aspects were addressed revealed the predominance of the original instrumental role given to Antarctic science over the physical presence in the region, considered quintessential for participating in the political decision-making process.

As a final note, it is necessary to state that the flexible approach of Brazil's policy for Antarctica is what allowed it to adapt rapidly to the changes experienced over that period in the international system, the Antarctic regime and the domestic political landscape. Also, it is important to highlight that the difficulties experienced by the PROANTAR were not exclusive to the programme, but a product of the economic hardships and political instabilities of the country. Thanks to the advancements of the programme and the commitment of a number of individuals, the PROANTAR was not only able to overcome those difficulties, but also to evolve into a more scientifically driven programme.

References

Alvarez, C. E. de. (1995). *Arquitetura na Antártica: Ênfase nas edificações brasileiras em madeira*. University of São Paulo. Retrieved from http://lpp.ufes.br/arquitetura-na-antártica-ênfase-nas-edificações-brasileiras-em-madeira

Avaliação preliminar do Programa Antártico Brasileiro. (2006). Brasília.

Bermejo García, R., & Bou Franch, V. (1993). La celebración del protocolo al tratado Antártico sobre protección del medio ambiente: orígenes y negociaciones. *Anuario Español de Derecho Internacional*, *9*, 155–211. https://doi.org/10.15581/010.9.155-211

Blay, S. K. N., & Tsamenyi, B. M. (1990). Australia and the convention for the regulation of Antarctic Mineral Resource Activities (CRAMRA). *Polar Record*, *26*(158), 195–202. https://doi.org/10.1017/S0032247400011438

Brasil na Antártica: 25 anos de história. (2008).

Câmara, P., Barros-Platiau, A., Andrade, I., & Hillebrand, G. (2021). Brazil in Antarctica: 40 years of science. *Antarctic Science*, *33*(1), 30–38. https://doi.org/10.1017/S0954102020000449

Cardone, I. J. (2015). *As Posições Brasileiras No Sistema Do Tratado Antártico Com Ênfase Na Questão Ambiental*. Universidade Federal de Parana. http://hdl.handle.net/1884/38844

Child, J. (1988). *Antarctica and South American geopolitics: Frozen Lebensraum*. Praeger.

Colacrai, M. (2003). Reflexiones en torno al régimen antártico y las relaciones argentino-chilenas. *Cuadernos de Politica Exterior Argentina*. Rosario: Centro de Estudios en Relaciones Internacionales de Rosario (CERIR), (73), 1–36.

Colacrai, M. (2012). *Continuidades y cambios en la Política Antártica Argentina 1959–2001: conjugación de factores internos y externos*. Editorial Académica Española.

Dalto, F. A. S. (2019). Brazilian financial crisis in the 1980s: Historical precedent of an economy governed by financial interests. *Revista De Economia Contemporânea, 23*(3), 1–25. https://doi.org/10.1590/198055272332

Farman, J. C., Gardiner, B. G., & Shanklin, J. D. (1985). Large losses of total ozone in Antarctica. *Nature, 315*, 207–210.

Fernandes, E. (1992). Law, politics and environmental protection in Brazil. *Journal of Environmental Law, 4*(1), 41–56. https://doi.org/10.1093/jel/4.1.41

Ferreira, F. R. G. (2009). *O Sistema do Tratado da Antártica: evolução do regime e seu impacto na política externa brasileira*. Fundação Alexandre de Gusmão.

Fontes Diégues, F. M. (1987). Cinco anos de presença da Marinha na Antártida. *Revista Marítima Brasileira, 107*(10–12), 97–106.

Fowler, A. N. (2000). *COMNAP: The national managers in Antarctica*. American Literary Press.

Gandra, R. M. (2013). *Geopolítica Antártica no Limiar do Século XXI: a definição de um projeto estratégico-científico para o Brasil na Antártida*. Universidade Federal de Rio Grande do Sul.

Joyner, C. C. (1987). The Antarctic minerals negotiating process. *The American Journal of International Law, 81*(4), 888–905.

Mattos, L. F. de. (2014). A Inclusão da Antártida no Conceito de Entorno Estratégico Brasileiro. *Revista Da Escola de Guerra Naval, 20*(1), 165–192. http://www.egn.mb/arquivos/revistaEgn/nova-revista/revista-20n1/index.htm

Modelo Brasileiro para o Desenvolvimento das Atividades Voltadas para os Recursos do Mar. (1998). *Revista Marítima Brasileira, 118*(7–9), 25–56.

Moneta, C. J. (1986). La Antártida y el Atlántico Sur en el sistema internacional: Alternativas de conflicto y vías de cooperación. *Revista De Estudios Internacionales, 7*(3), 837–853.

Morgan, C. (1988). Australia reconsiders claims to Antarctic mineral resources. *Nature, 336*(6198), 417–417. https://doi.org/10.1038/336417a0

Marítimo, N. (1994). NApOc Ary Rongel. *Revista Marítima Brasileira, 114*(7–9), 261–263.

Noticiário Marítimo. (2003). *Revista Marítima Brasileira, 123*(7–9), 279–320.

Nunes, M. R. (2005). *O Brasil na Antártica*. Régis Aló.

Soares Guimarães, L. F. (1986). Um Brasil Ignorado: O espaço marinho nacional. *Revista Marítima Brasileira, 106*(7–9), 121–140.

Villa, R. A. D. (2004). *A Antártida no Sistema Internacional*. Hucitec.

Williamson, J. (2009). A Short History of the Washington Consensus. *Law and Business Review o the Americas, 15*(1), 7–24.

Archival Sources

Atividades da área de política externa da 1ra Subchefia da SG/CSN. (1987, May 25). [Report on Foreign Affairs]. Dossié Política Externa, pp. 67–73 (BR/DFANBSB/N8.0/PSN/EST.96), Arquivos Nacionais, Brazil.

McTaggart, D. (1991, April 15). Re: the protection of the Antarctic environment [Letter to Francisco Rezek]. Antártida.—Dossiê, pp. 703–704. (BR DFANBSB Z4.REX.IPS.26), Arquivo Nacional, Brazil.

Newspapers and Other Sources

Antarctic Resources: effects of mineral exploration. (1972, November 2). ANT/44 (ATCM VII/WP44), Antarctic Treaty Secretariat.

Antarctic Specially Managed Area—ASMA da Baia do Almirantado. (1994, June 7). RAPAL V: Doc. 38/94. RAPAL.

Ata da 161 Sessão Ordinária. (2006, April 27). [Proceedings]. Comissão Interministerial para os Recursos do Mar, Brazil.

Bird Denuncia destruição da Amazônia. (1988, September 22). *Jornal do Brasil* (Rio de Janeiro, Brazil), *98*(167), 12.

Bird suspendió recursos de US$375 milhões ao Brasil. (1988, September 24). *Jornal do Brasil* (Rio de Janeiro, Brazil), *98*(169), 14.

Brasil completa 12 anos na Antártida em meio a crise. (1995, January 15). *Jornal do Brasil* (Rio de Janeiro, Brazil), *104*(282), 9.

Brasil está fora da Antártida. (1992, December 16). *Tribuna da Imprensa* (Rio de Janeiro, Brazil), *43*(13.070), 11.

Comitê Nacional de Pesquisas Antárticas é reativado. (2004, June 16). São Paulo, Agência FAPESP. Source: https://agencia.fapesp.br/comite-nacional-de-pesquisas-antarticas-e-reativado/1967/. Accessed on: 10 December 2020.

Comunicado Conjunto Brasileiro-Soviético sobre a Visita Oficial do Ministro dos Negocios Estrangeiros da URSS ao Brasil. (1987, September 30). Source: https://aplicacao.itamaraty.gov.br/ApiConcordia/Documento/download/7520. Accessed on: 10 December 2020.

Comunicado Conjunto Brasileiro-Soviético. (1988, October 19). Source: https://aplicacao.itamaraty.gov.br/ApiConcordia/Documento/download/7329. Accessed on: 10 December 2020.

Comunicado Conjunto entre o Governo da República Federativa do Brasil e o Governo da República Argentina. (1990, July 6). Brazil.

Costeau quer bloquear mineração na Antártica. (1989, October 10). *Jornal do Brasil* (Rio de Janeiro, Brazil), *99*(185), 7.

Declaração Conjunta entre o Governo da República Federativa do Brasil e o Governo da República do Chile. (1990, July 36). Brazil.

Declaración de Montevideo. (1991, July 1991). RAPAL. Source: http://www.rapal.org.ar/RAPAL2.PDF. Accessed on: 10 December 2020.

Declaración Final. (1990, June). Reunión de Administradores de Programas Nacionales Antárticos Latinoamericanos, Buenos Aires, Junio de 1990. RAPAL. Retrieved from: http://www.rapal.org.ar/RAPAL1.RTF. Accessed on: 10 December 2020.

Decreto Legislativo N° 37. (1990, October 26). Brazil.

Decreto Legislativo N° 88. (1995, June 6). Brazil.

Decreto N° 66. (1991, March 18). Brazil.

Decreto N° 123. (1991, May 20). Brazil.

Decreto N° 1.791. (1996, January 15). Brazil.

Decreto N° 91.146. (1985, March 15). Brazil.

Decreto N° 92.522. (1986, April 7). II Plano Setorial para os Recursos do Mar. Brazil.

Decreto N° 92.878 (1986, June 30). Brazil.

Decreto N° 93.935. (1987a, January 15). Brazil.

Decreto N° 94.401. (1987b, June 3). Brazil.

Decreto N° 94.679. (1987c, July 24). Brazil.

Expedição à Antártida pode ser militarizada. (1991, August 25). *Tribuna da Imprensa* (Rio de Janeiro, Brazil), *42*(12.787), 5.

Exposição de motivos DNU/DPB/DAI/ SAL/106.10 (D). (1984, May 8). Diario do Congresso Nacional, Seção 1, ANO XL, N° 006, 8 March 1985, p.0528. Camara dos Deputados, Brazil.

Item 8: Mineral resources. Effects of prospecting for minerals. (1972, October 31). ANT/37 (ATCM VII/WP37). Antarctic Treaty Secretariat.

Joint Statement on International Environment Issues Agreed by Prime Ministers Hawke and Rocard, Canberra. (1989, August 18). (ATCM XV/ BP01), Antarctic Treaty Secretariat.

Mensagem N° 231 de 1993. (Do Poder Executivo) (1993, April 30). Diário do Congresso Nacional (Seção I), *XLVII*(091), 1119. Cámara dos Deputados (Brazil).

Opening address by Mr. Henrique R. Valle, Leader of the Delegation of Brazil. (1989, October). Final Report of the Fifteenth Antarctic Treaty Consultative Meeting, Paris 9–20, pp. 135–137. Antarctic Treaty Secretariat.

Opening address by the Head of the Delegation of Brazil, H.E. Ambassador Henrique R. Valle. (1991, October 9). ATCM XVI/IP51 [info 051]. Antarctic Treaty Secretariat.

Opening statement by Mr. Henrique R. Valle, Head of the Brazilian Delegation. (1990, December). Interim Report of the Eleventh Antarctic Treaty Special Consultative Meeting, Viña del Mar, 19 November–6 December 1990, p. 33. Antarctic Treaty Secretariat.

Política de Defesa Nacional. (1996). Brazil.

Proantar precisa de verbas. Brasil é contra a exploração mineral da Antártica. (1991, February 17). *Jornal do Brasil* (Rio de Janeiro, Brazil), *100*(311), 20.
Proposed Amendments to Mr.17-Rev.III Arts. 2, 14, 22, 26, 29, 31, 46, 51. (1987, May). SATCM IV-10/WP18. Antarctic Treaty Secretariat.
Report of the fifth meeting of the commission, Hobart, Australia. (1985, September). CCAMLR V, Commission for the Conservation of Antarctic Marine Living Resources.
Report of the third meeting of the commission, Hobart, Australia. (1984, September). CCAMLR III, Commission for the Conservation of Antarctic Marine Living Resources.
Resolution 43/83. (1988, December 7). A/RES/43/83. Resolutions and Decisions adopted by the General Assembly during its 43rd Session Volume I, 20 September–22 December 1988, pp. 103–104. United Nations.
Resolution 43/196. (1988, December 20). A/RES/43/196. Resolutions and Decisions adopted by the General Assembly during its 43rd Session Volume I, 20 September–22 December 1988, pp. 147–148. United Nations.
Statement by Mr. Antonio Guerreiro, Head of Delegation, Brazil. (1988, June). Final Report of the Fourth Special Antarctic Treaty Consultative Meeting on Antarctic Mineral Resources, Wellington, 2 June 1988, pp. 26–27 (SATCM IV-12), Antarctic Treaty Secretariat.
Statement by the Representative of the Government of Brazil. (1985, September). Report of the fourth meeting of the commission, Hobart, Australia, 2–13 September 1985, ANNEX F. CCAMLR IV: Commission for the Conservation of Antarctic Marine Living Resources.

CHAPTER 5

Science and Environment at the Central Stage

5.1 The Restructure of Brazil's Antarctic Programme: From the Research Networks to the 4th International Polar Year

The beginning of the twenty-first century marked a restructuring of the Brazilian Antarctic programme. The last two years of Cardoso's Presidency (2001–2002) saw a reframing of Antarctic research through the active involvement of the Ministry of Environment and a redefinition of its scientific aims, following the Protocol of Environmental Protection. At the same time, the first evidence of the passage of time began to be felt in the Brazilian Antarctic infrastructure, which made it necessary to renew the commitment of the government to invest in the programme.

The course of almost twenty years of the Brazilian programme began to make the severe weather conditions of the Antarctic visible. By 2001, the unplanned growth of the *Estação Antártica Comandante Ferraz* (EACF) evidenced issues regarding the efficiency and maintenance of its infrastructure, which could impact negatively on the safety of operations and the, up until then, exemplary environmental performance. Of particular importance was the corrosion of the structure of the station as well in the oil tanks. Acknowledging these problems and the potential need of further growth and compliance of the Environmental Protocol, the PROANTAR began studies on to the elaboration of a master plan to review the station infrastructure and the need for adaptations. The master

I. J. Cardone, *The Antarctic Politics of Brazil*,
https://doi.org/10.1007/978-3-030-80161-8_5

plan was centred on three aspects related to the Brazilian presence in the continent: comfort and efficiency; acoustics; and corrosion.[1]

Additionally, some unintended developments that changed the Brazilian programme in the years to come were taking place. By the end of Cardoso's Presidency, the Ministry of Environment started to show an interest in a more active involvement in the programme. Key to that involvement was the figure of Dr Tânia Brito, by then coordinator of the Environmental Assessment Group of the subcommittee of the PROANTAR in the CIRM. Thanks to the efforts of Dr Brito and her active involvement in the programme, in 2002, the Ministry of the Environment decided to promote the development of an issue-specific research project[2] in line with the increasing interest of the global community and the Scientific Committee on Antarctic Research (SCAR) in climate change: Environmental Change in Antarctica: local and global impacts.[3]

The issue-specific project was designed to interlink different research groups and disciplines in a single research framework. To accomplish this, the project was organized in two research networks: (1) Global Environmental Impacts: aimed to study the impacts of global and regional environmental changes in Antarctica, it included twelve research teams divided into seven thematic groups, including sixteen Brazilian institutions and some seventeen other international partners from seven different nations; and (2) Environmental Monitoring: designed to establish the impacts of human activity in Admiralty Bay (*Lasserre*) of King George Island (*25 de Mayo/Vaterloo*)—which, since 1996, had been established as an Antarctic Special Managed Area thanks to the initiative of Brazil and Poland—it was composed of fifteen research teams from eight Brazilian institutions.[4]

The timing of the involvement of the Ministry of Environment could not have been better, as the second network would assist and reinforce the work of the team that was developing the master plan for the EACF, and the financial resources added to the programme through the

[1] Alvarez et al. (2007).

[2] The PROANTAR called this an 'induced' project. It refers to projects that fell inside a predefined research issue. In this chapter, this type of call is termed as 'issue-specific.'

[3] MMA, *O Brasil e o Meio Ambiente Antártico / Ministério do Meio Ambiente* (2007).

[4] Iniciativas Ambientais Brasileiras (2002).

project aided the PROANTAR in times of financial strain. In 2002, the resources for logistic activities were cut, which undermined the continuity of ongoing scientific projects and made the CIRM Secretariat consider the possibility of limiting the activities undertaken by the XXI Brazilian Antarctic Expedition (OPERANTAR).[5] The research networks promoted by the Ministry of Environment added 1 million reais (about USD 350,000) to the logistics and 1.8 millions reais (about USD 635,000) for research activities, enabling the materialization of the OPERANTAR XXI, which included thirty research projects in Antarctica, twenty-six from the research networks and four ongoing projects.[6]

The end of Cardoso's Presidency and the advent of 'Lula' da Silva's Presidency, of the Worker's Party, in January 2003, did not result in any significant change in Antarctic policy. The financial problems experienced by the Navy to finance the PROANTAR's logistics, the National Research Council's (*Conselho Nacional de Desenvolvimento Científico e Tecnológico*—CNPq) difficulties to provide funding for Antarctic research and the general limitations to promote Brazilian representatives' participation in international forums motivated a strong lobby that resulted in the constitution, by a group of the Federal Representatives, of the *Frente Parlamentar de Apoio ao Programa Antártico Brasileiro* (Congressional Group in Support of the Brazilian Antarctic Programme) in 2003.[7] When announcing the creation of the group, the Representative Josias Quintal lamented that:

> ...the PROANTAR has confronted great challenges to its continuity, due to the increasing cuts on funding, which resulted on the lack of an adequate response to logistical and scientific needs, as would be appropriate. And, as even the support to a minimum of research projects has been difficult, the programme has been subject to severe limitations. Added to that, it is necessary to consider the cuts experienced in the funding of the Navy -which has been taking place since 1990—, what rise the anxiety of everyone in what respect the prospects of continuity of the work of the

[5] 148ª Sessão Ordinária, Item 6.1.2.2, lines 564–583.

[6] 149ª Sessão Ordinária, Item 6.1.2.5, lines 436–442.

[7] Many sources, including Ferreira (2009, p. 153), locate the creation of the group in 2007. Whether there was an announcement of the creation of such group in the Brazilian Senate in 2007, it seems to have been merely the renewal of the pre-existent group in the following congressional period. The group with that designation existed since 2003 as can be seen in the references provided.

> programme. That situation, added to being disturbing, is reprehensible, given the strategic importance of this program, developed and maintained thanks to the idealism, generosity and nationalism of a group of Brazilians and their institutions.[8]

The creation of the Congressional Group provided political influence on the programme, especially important when obtaining financial resources, and was fundamental in the subsequent years to obtain amendments to the national budget to complement the funding of Antarctic logistic, infrastructure and scientific activities.[9] As well as the challenges that the programme confronted, the Congressional Group added another layer of influence with direct reach to the decision-making process in the allocation of resources. However, whether the Group was key in making the programme sustainable in the long run, it also established a political dependence that submitted that continuity to the political and economic fluctuations. In particular, the influence depended on the formation of the group on each congressional period. More importantly, the continued need for intervention of such group in what was regarded the budget allocation of funds to the programme through amendments made the relative lack of political priority of the programme apparent.[10]

While the programme was gaining political support in Brasília, in Antarctica, the shelter Padre Balduíno Rambo, deactivated in 1990 and used only for emergencies, collapsed due to corrosion.[11] The collapse damaged the shelter beyond repair and the PROANTAR decided to dismantle and remove it.

Added to the financial problems and the deterioration of the infrastructure of the PROANTAR, other limitations in terms of regulation and implementation of the Antarctic regime's rules started to become apparent. In November, Amyr Klink, a famous Brazilian sailor and explorer, informed the CIRM of his plans to circumnavigate the Antarctic

[8] Diário da Câmara dos Deputados (2003, p. 67110, own translation).

[9] Cardone (2014, 2015).

[10] Ferreira considers the creation of the Group an explicit recognition of the political importance of the programme (Ferreira, 2009, pp. 153–154). However, the acquisition of political prominence by the programme would make the need of such an intervention innocuous and actually reveals the point at which the programme was not considered a priority in many governmental circles.

[11] Resultado da Avaliação Ambiental (2005).

continent in a sailing vessel. The absence of any regulation regarding the authorization from the Brazilian establishment in charge of Brazilian activities in Antarctica, namely the CIRM, deprived the authorities of any legal mechanism to require previous permission from the Environmental Assessment Group of the PROANTAR that could ensure that the necessary environmental precautions were adopted in compliance with those required by the Protocol of Environmental Protection.[12] This limitation was not new, but it showed the lack of any control over tourist activities, a concern that Brazil has raised in all Antarctic forums since then.

A stable economy added to international circumstances that resulted in the increased relative price of commodities presented a unique opportunity for President 'Lula' da Silva. Taking advantage of the resultant economic growth and a favourable balance of trade, the President accumulated important international monetary reserves, promoted social policies focused on the lower classes and favoured the industrial sector through incentives in the domestic market and an active a programme of heavy public investment, which included public–private partnerships. The improvement in the economy and the progressive policy of 'Lula' da Silva's administration led to increased funding for areas such as science and technology, including universities and scientific agencies. That did not result in an immediate amelioration of the financial situation of the PROANTAR but did offer the context in which such improvement was made possible.

The year of 2004 would be relevant due to the resume of the activities of the Brazilian Antarctic scientific body, the *Comitê Nacional de Pesquisas Antárticas* (National Committee on Antarctic Research—CONAPA), which met in June after six years of inactivity. The reactivation of the CONAPA was considered essential to have a more integrated representation on the SCAR and increase the international visibility of the Brazilian Antarctic science, taking advantage of the work of the two research networks and favouring international partnerships.[13] In addition, the announcement of the 4th International Polar Year (IPY) 2007–2008 by the end of 2004 gave an additional incentive to the programme.

In Antarctica, the studies and works to adapt the structures continued, despite the lack of resources. Between November and December

[12] 154ª Sessão Ordinária (2003, Item 6.3, lines 340–356).

[13] Comitê Nacional de Pesquisas (2004).

2004, the shelter Padre Balduíno Rambo was completely dismantled with a minimum residual environmental impact, displaying the country's commitment with the environmental standards established by the Protocol of Environmental Protection.[14] For its part, the studies of the EACF showed the structures of the modules and oil tanks showed significant signs of corrosion that made the need for repairs and adaptations imperative.[15] Added to that, the CIRM started to study different options to replace the communication system of the station, to reduce the dependence on Chile, which provided the system up until that time.[16]

Simultaneously, the glaciologist Jefferson Simões, of the Federal University of Rio Grande do Sul, was participating in a joint expedition with the Chilean Antarctic Programme to traverse from the Chilean Parodi Station[17] to the South Pole. The feat was accomplished on 30 November 2004, making Simões de first Brazilian national to reach the South Pole by land. On their traverse back to the station, the expedition took important ice-core samples for study.

In the year 2005, despite the pervasive financial difficulties, the plans for renovating the EACF and the involvement of Brazil in the upcoming International Polar Year advanced. However, the developments of the Antarctic Treaty System (ATS) would be received with some dissatisfaction by Brazil. The adoption of the Annex VI of Liability at the XXVIII ATCM in Stockholm was received with contempt by the Brazilian government, which considered that, as approved, the Annex did not provide much definition on several aspects that would be needed to its effective implementation. To the Brazilians, the annex was undefined in how to apply the conditions to non-parties, how to determine what would be considered as a 'significative' impact, how defined is the party responsible for the activities and by which mechanisms the measures adopted by the Protocol would be internalized by each country as a way to enforce its compliance, as well as other issues.[18]

Following what was expressed in the previous years, Brazil was increasingly concerned about the impact that activities in Antarctica could have,

[14] Resultado da Avaliação Ambiental (2005).

[15] Ata 157ª Sessão Ordinária (2004, Item 6.3.1, lines 754–758).

[16] Ibid., Item 6.2.2, lines 716–721.

[17] Since 2013 renamed Union Glacier Station.

[18] Considerações e Implicações (2005).

not only on the pristine continent of Antarctica, but also indirectly on the South American region and, in particular, over Brazil. The two activities that, by that time, aroused the highest concerns in Brazil were fishing in the Antarctic and unregulated tourism.

These concerns and the increased political leverage gained in recent years by the programme had favoured the explicit acknowledgement of the country's interest in Antarctica. During the second round of the meetings of the Ministers of Science and Technology of the group of India, Brazil and South Africa, on 9 June 2005, the Antarctic region was acknowledged as an area in which those countries should pursue cooperation. Also, the revision of that year of the National Defence Policy reedited the reference to Antarctica by defining within its objectives that of: '…actively participate in the decision-making processes for the Antarctic region.'[19] Whether the new National Defence Policy did not include Antarctica within the region of strategic interest—the so-called strategic surrounding—the aim of deciding on the futures of the continent showed a continuity in how the country linked the decisions on Antarctica to matters of national security.

Advancing with the project of becoming involved in the IPY, the year of 2005 was relevant for the Brazilian Antarctic science. On 17 June 2005, the Ministry of Science and Technology established the working group *Agenda Antártica* (Antarctic Agenda), to establish a five-year strategic plan for the years 2006–2010 that would include: Brazil's participation in the IPY, the logistical and financial needs for Antarctic research, the possible integration of activities and funding sources and the draft of proposals to be included in the pluriannual national budget.[20]

The report of the working group proposed updating and strengthening the CONAPA. On a domestic level, it was suggested the CONAPA should be dynamized to promote a more active national Antarctic scientific policy and its composition be reviewed to represent the incorporation of new actors within the PROANTAR. In the international realm, it was anticipated a more active participation within the SCAR and its involvement in the organizing committee of the IPY. The report stressed the need to advance actions in 2006 in order to guarantee an effective participation of Brazil in the international event and called for integrated

[19] Decreto N° 5484 (2005, subitem XXII, item 7.1., own translation).

[20] Relatório Final: GT (2005).

actions that would favour international collaboration and efficiency in the use of resources. Also, it recommended the elaboration of a pluri-annual plan for the PROANTAR to promote medium-term objectives and to support the creation of research centres specialized in Antarctic fields of research. Pursuing those objectives, the revision of the Brazilian Antarctic Policy of 1987 was also recommended; the support of integrated inter-disciplinary, inter-institutional and international Antarctic research projects; the promotion of specialization of human resources in Antarctic fields of expertise; and the expansion of the area of research, including Inner Antarctica, among others. Following the tradition of linking the Antarctic interest to the Brazilian mainland, it was also stressed that support of Antarctic research should be granted specifically to those fields in which links with the Brazilian territory could be established.

Additionally, between the 25 and 29 July 2005, the 9th Biology Symposium was held in Curitiba, Parana. It was the first symposium of the kind to be held in South America and was considered a great success by the international scientific community.[21] It was promoted by the prominent Brazilian biologist Edith Fanta, one of the pioneer biological researchers of the PROANTAR, which was the SCAR representative to the Convention on the Conservation of Antarctic Marine Living Resources (CCAMLR) and part of the Joint Committee of the IPY since 2004. To add to Fanta's international recognition, that year, she was elected Chair of the Scientific Committee of the CCAMLR, a position she held until her premature death in 2008.[22] Jointly with the secretary and president of the SCAR held by Dr Antônio Rocha-Campos,[23] the designation of Dr Fanta on the highest scientific advisory committee for living resources was a formidable distinction for Brazilian Antarctic science and a display of the attained level of scientific development and international integration.

This coincided with the inclusion of the Special Secretary of Aquaculture and Fishing, dependent on the Presidency, within the subcommittee

[21] *Scientific Committee on Antarctic Research Bulletin No. 161, December 2006* (2006, p. 23).

[22] Krupnik et al. (2011, p. 636) and Ferreira (2009, p. 140). Ferreira mistakes Fanta's role in the Joint Committee with the Organizing Committee.

[23] As seen previously, Rocha-Campos occupied those post between 1986–1992 and 1994–1998, respectively.

of the PROANTAR.[24] It is unclear if such incorporation was motivated by the active involvement in the Brazilian programme with the CCAMLR, but both events certainly signalled an increased interest in the issue of Antarctic marine living resources. This is particularly important when considering that Brazil had joined the other South American countries in pursuing the need to protect the Antarctic related ecosystems and that fisheries in the South Atlantic and South Pacific were strongly affected by changes in the Southern Ocean.

After three years of technical surveys of the conditions of the program's Antarctic infrastructure, by 2005 the master plan for the EACF was concluded. The conditions of the station and the oil tanks had raised safety concerns within the Navy, which considered the refitting of the station as essential to ensure successful involvement of the PROANTAR on the IPY in conditions that could guarantee the safety of personnel.[25] The plan included important modifications and adaptations to be implemented in a three-year period, aimed at improving the safety, comfort and environmental efficiency of the operations.[26] With the master plan concluded, in September of that year, the work of refitting the EACF started and were conducted directly by the Brazilian Navy personnel. Additionally, and following the studies implemented the previous year, the installation of a new national system of communication for the EACF started.

By the time Brazil participated on the XVI *Reunião de Administradores de Programas Antárticos Latinoamericanos* (Meeting of Administrators of Latin American Antarctic Programmes—RAPAL), the first signs of recovery of the programme started to appear. Along the last years, participation of Brazil on the RAPAL meetings was very limited, making it difficult to follow and participate in all discussions and subgroups, as the issues and specialties discussed in the meetings expanded.[27] But the Brazilian delegation to the XVI RAPAL was more numerous and specialized.[28]

[24] Ata 160ª Sessão Ordinária (2005, Item 3.2, lines 162–166).

[25] Ata 160ª Sessão Ordinária (2005, Item 6.5.1, lines 745–766).

[26] Conceitos e Critérios Adotados (2005).

[27] 149ª Sessão Ordinária (2002, Item 6.1.3.4, lines 466–473).

[28] 154ª Sessão Ordinária (2003, Item 6.2.1, lines 316–320).

Brazil's delegation to the XVI RAPAL included representatives of the CIRM Secretariat, the Brazilian Air Force, the Ministry of Science and Technology, the Ministry of Environment, the CNPq and four universities—including the two research networks. The Brazilian delegation promoted the idea of coordinating efforts for the upcoming IPY and advanced the idea of constituting a Latin-American IPY programme, including a census of Antarctic marine life in South America. This proposal was in line with the importance that the marine living resource was gaining in Brazil thanks to the work of Dr Fanta and her group.[29]

Notwithstanding, the Navy continued to experience serious problems of covering the programme's costs of logistics and maintenance of the infrastructure.[30] The situation was so critical that the Ministry of Foreign Affairs expressed its concerns about the possibility of Brazil downgrading its status within the ATS. The budget for the necessary adaptations to the Ary Rongel, the aircrafts and the EACF amounted an approximate 18 million reais (about USD 7.7 million), following the three-year plan.[31] That added to the three million reais (about USD 1.3 million) that regular operations required annually. With the funds of the Navy exhausted by the end of 2005, during the 160 Ordinary Meeting of the CIRM, its coordinator, Admiral Roberto de Guimarães Carvalho, recalled the other ministries that 'The PROANTAR is a State endeavour, not being exclusive of the Navy. Thus, it cannot be only sustained by the Brazilian Navy budget, being necessary the funding from other sources.'[32]

Despite all the financial difficulties, in March 2006, the communication system installed by the Brazilian company Telemar was in operation. The new system not only provided autonomy from Chile—something relevant when considering that it connected the Antarctic and the Brazilian territory without the intervention of third parties—but also improved the communications. The new system included access to the internet, six VoIP telecommunication lines, reception of TV signal from the Globo

[29] O Censo de Vida Marinha Antártica: Iniciativa para o Ano Polar Internacional 2007/08 (2005) and Proposta Brasileira (2005).

[30] Ata 160ª Sessão Ordinária (2005, Item 6.5, lines 699–705).

[31] Ibid., Item 6.5.1, lines 721–742.

[32] Ibid., Item 6.5.1, lines 743–745, own translation.

Network—the most popular TV channel in Brazil—and data connection with the *Ary Rongel*, when in the proximities of the station.[33]

By April, the financial strains of the programme received its first relief with the concession of an extraordinary fund of 10 million reais (about USD 4.6 million) for the refitting of the EACF.[34] The added funding made the most urgent task of renovating the Antarctic infrastructure viable, even when additional funding for the *Ary Rongel*, the aircrafts and participation on the different international forums was still wanting. Additionally, in pursuance of alternative sources of funding, an agreement with the *Fundação de Estudos do Mar* (Studies of the Sea Foundation) was signed to facilitate the raising of private funds to finance scientific research in Antarctica.[35]

Institutionally, due to the importance that the Antarctic tourist industry was acquiring and the interest that Brazil had raised on the issue in recent times, the Ministry of Tourism was included in the subcommittee of the PROANTAR. The Ministry was already included in the *Comisão Nacional para Assuntos Antárticos* (National Commission for Antarctic Issues—CONANTAR) in 2000,[36] but as the commission was inactive since 1995, that inclusion had been ineffective until that date. It was expected that the upcoming IPY would increase the interest in Antarctic tourism and the transportation of scientific personnel to and from the continent. Thus, the inclusion of the Ministry of Tourism was seen as key to seize the potential benefits that such growth of Antarctic travel could bring to cities that operated as stops and service providers.

In view of the upcoming IPY and the challenges that it posed to the Brazilian Antarctic programme, all the institutional structure of the PROANTAR was activated. In May 2006 and after ten years of inactivity, the CONANTAR met again, thanks to the lobby of the Ministry of Foreign Affairs and the CIRM.[37] The CONAPA also met, proposing extensive involvement in the international scientific event, including all the ministries already involved in the programme and adding the Ministry of Education and Culture. This inclusion followed the strong educational

[33] Ata 161ª Sessão Ordinária (2006, Item 6.8, lines 944–968).

[34] Ibid., Item 6.3, lines 855–863.

[35] Ibid., Item 8.2, lines 1086–1094.

[36] Decreto N° 3.416 (2000).

[37] Ata 161ª Sessão Ordinária (2006, Item 6.5, lines 907–913).

aspect promoted by the IPY organizing committee and was oriented to the promotion of the programme and the Antarctic consciousness between the general public of Brazil. The inclusion of the Ministry of Education and Culture in the subcommittee of the PROANTAR in September of that year cemented the involvement of the educational element within the programme.[38]

The changes in the ATS, the domestic setting and the upcoming IPY called for a reformulation of the objectives of the Brazilian Antarctic programme. Much had changed in terms of the research developed in the country, the fields of interest and the foreseen areas for development in Antarctic science since Brazil's first involvement in Antarctica. In view of the need to adapt the programme to those circumstances, in December 2006, the CIRM approved a new version of the PROANTAR.

The new PROANTAR did not significantly alter the main political guidelines and objectives. It was designed following the 1987 POLANTAR, the resolutions of the CONANTAR, the recommendations of the CONAPA, the resolutions of the CIRM and the principles and objectives of the Brazilian foreign policy, condensing all the previous legislation relative to the institutional design.[39] In compliance with the conditions established by the Protocol of Environmental Protection, it established that all Antarctic activities, including scientific research, tourism, education, dissemination and artistic expressions, should be undertaken reducing the impact over the Antarctic environment to the maximum possible extent, and should be submitted for consideration of the subcommittee of the PROANTAR for the analysis of its different aspects in each sub-group.

In terms of its objectives, those of the 2006 PROANTAR aimed to promote a diverse, high-quality and relevant Antarctic research; contribute to Brazil's participation in all international Antarctic forums; demonstrate Brazil's interest in the region by promoting its presence in the continent; promote international cooperation following the national objectives; promote the conservation of the Antarctic environment and its related ecosystems, including to collaborate in any international effort to that end; develop technologies that could reduce the impact and improve the safety and comfort of operations; promote the specialization of human

[38] Resolução N° 7 (2006).

[39] Programa Antártico Brasileiro (2006, p. 5).

resources in Antarctic fields of research; and prioritize safety of operations and comfort of those activities to be undertaken in Antarctica. In general, the reformulated objectives did not innovate considerably, but located more emphasis on aspects related to scientific research, especially in those fields that were developing at that time. Additionally, incorporating some contemporary concerns, it included tourism and education between its considerations.

In what respect the way in which the programme was structured, the 2006 PROANTAR added two sub-programmes to the original five. As finally approved, the PROANTAR was composed of three science-related sub-programmes: physical sciences, geo sciences and life sciences; one technological: technological development; and three related to support activities: environment; education, training and awareness; and logistics.[40] The new programme did not bring any relevant change to the scientific-field, changing the name of the atmospheric sciences by the more encompassing of physical sciences, but incorporated the sub-programme of environment, needed to bring support to the responsibilities assumed as part of the Protocol of Environmental Protection; and technological development, which could accrue the benefits of the research that was being undertaken in infrastructure and operations.

Following the long tradition of promoting research that could be presented as having a direct or indirect impact on the Brazilian mainland, in general, all sub-programmes were established with a view to a general comprehension of global phenomena but linked to its potentialities related to economic interests within Brazil. The physical sciences, for instance, were praised for their potential to, indirectly, improve the agricultural activities by better weather forecasting, which would also improve the safety of Antarctic operations. Also, the physical relations between the pole and the tropical region were established by linking ionospheric phenomenon in the tropic with precipitation of solar particles in the pole, as well as the different interchanges between the Southern Ocean and the South Atlantic.[41] Moreover, the sub-programme of geosciences was stressed by the connection between the geological, geophysical and glaciological aspects of Antarctica and phenomenon that occurs in Brazil. Although the sub-programme of life sciences appeared as specific to the

[40] Ibid., p. 14.

[41] Ibid., pp. 14–15.

Antarctic region, it also included the possibility of '...enabling future activities of renewable resources rational exploitation.'[42] By its part, the sub-programme of technological development main aim was to improve the efficiency and environmental performance of Brazilian operations, including, as secondary objective, the implementation of technologies developed in Antarctica within the Brazilian territory. Similarly, the sub-programme of environment, mainly directed to ensure the compliance of the Brazilian operation with the environmental regulations in Antarctica, added as secondary objective that of following the development of tourist activities and to study the implementation of knowledge developed in Antarctica into the Brazilian context. The only exceptions to these considerations were the sub-programme of education, training and awareness, aimed at promoting the specialization of human resources and the connection between Brazil and Antarctica within the public imagination, and the logistic sub-programme, both of which were essential to make all others possible.

In 2007, 'Lula' da Silva assumed his second term as President of Brazil after a categorical victory at the polls. The success of his first Presidency, reflected in the polls and in international opinion, led to the continuance of his main policies, including the scientific and technological development. The heavy investment in development and the promotion of the international profile of Brazil was something that the Brazilian Antarctic programme could exploit, in particular in front of the 25th anniversary of the programme and the upcoming IPY. But the programme still lacked a stable source of funding that would allow it to meet its long-term financial needs. In February 2007 and following that necessity, the Brazilian representatives reinstalled the Congressional Group, this time under the name of *Frente Parlamentar Mista de Apoio ao Programa Antártico Brasileiro* (Bicameral Congressional Group in Support of the Brazilian Antarctic Programme).

On 1 March 2007, the 4th IPY was launched.Thanks to the many efforts of multiple actors during the previous years, Brazil's participation on the international scientific event was a success, with ten projects amounting more than 9 million reais (about USD 4.25 million) for research only.[43] If it is considered that total resources allocated to the

[42] Ibid., p. 16.

[43] Informe Final (2007, p. 3).

programme went from almost 12 millions reais in 2004 (about USD 3.9 million) up to 50 millions in 2006 (about USD 23 million),[44] it is possible to assess the impact that the efforts of the scientific community, the Ministry of Environment, the Navy and the Congressional Group, coupled with the interest generated by the IPY, had in advancing a more proactive Antarctic policy. In just a few years the Antarctic programme was able to renew its scientific programme, refit its infrastructure and take the initiative in environmental monitoring and conservation through leadership of the monitoring of the Antarctic Specially Managed Area of Admiralty Bay. In addition, the initiative that Brazil took in environmental conservation allowed that, in the 10th Committee of Environmental Protection that took place at the same time as the XXX ATCM in New Delhi, India, Dr Tânia Brito, of the Brazilian Ministry of Environment, was elected coordinator of an informal working group established to review the management plans for special protected and managed areas.

Along with those efforts, Brazil also intensified its Antarctic cooperation with its Latin-American partners. Between 26 and 29 September 2007, Brazil hosted the XVIII RAPAL in Brasília. The country's contributions to the meeting were focused on the environmental issue, including the achievements of the study of the cumulative impacts of human presence in Admiralty Bay and the recent works in adaptation of the Brazilian infrastructure. Furthering South American cooperation in Antarctic research, the following month, the National Research Council approved the South American Network on Antarctic Marine Biodiversity project. The proposal was built on the work of the census of Antarctic marine life in South America and the cooperation between Argentina, Brazil, Chile, Ecuador, Peru, Uruguay and Venezuela to structure a research network in Antarctic marine biodiversity, promoting the interchange of information and the interaction between researchers and administrators.[45]

Brazil's considerable participation in the IPY, the increased funding of the programme and the active regional cooperative attitude of Brazil was a

[44] The actual values are: 11,992,803.64 for 2004; 13,188,511.00 for 2005 and 50,243,924.40. Information retrieved from: Planilha de Custos 2004. CIRM 155, April 2004; Planilha de Custos 2005. CIRM 158, April 2005; and Planilha de Custos 2006. CIRM 162, May 2006. Source: https://www.marinha.mil.br/secirm/atas. Accessed on: 20 December 2020.

[45] South American Network (2008).

signal of new times for the PROANTAR. It was a golden era for Brazilian Antarctic science. However, these conditions depended on other factors, and they were not destined to last.

5.2 The Consolidation of the Scientific Profile of PROANTAR: The Golden years

Before the closing of the IPY in Antarctica, between 17 and 19 February 2008, President 'Lula' da Silva visited the EACF, commemorating the twenty-five years of the official arrival of Brazil on the continent. The visit of the President to the station and the *Ary Rongel* served to display the scientific works developed in the region and the results of the investments of the last years. But in contrast to the visit of President Collor in 1991, President 'Lula' da Silva used the visit to listen to the necessities of scientists and military personnel and offered to increase the resources to the programme in order to attain a 'qualitative leap' on the country's Antarctic research.[46] In this occasion, the Brazilian Navy Commander, Adm. Júlio Soares de Moura Neto, suggested the acquisition of a second polar vessel that would improve the research and logistic capabilities of the programme.[47]

This appeal resulted in the contract for the acquisition, by the end of April, of the vessel *Ocean Empress*, on what was considered by the press as the fastest answer to a Navy request.[48] The ship was being refitted in Bremerhaven, Germany, to resume its tasks as a Baltic Sea offshore oil operation support vessel, and was acquired from the owning company, the Russian ASK Subsea/Isis Viking Ltd. for 69 million reais (about USD 40 million).[49] It was christened *Almirante Maximiano* by the Brazilian Navy,[50] in recognition to the late Admiral Maximiano da Fonseca, the ardent promoter of the Brazilian Antarctic programme. The acquisition was a vital step towards increasing the logistical and scientific capabilities

[46] Lula parte para visita (2008) and Lula volta da Antártica (2008).

[47] Ata 168ª Sessão Ordinária (2008, Item 5.1, lines 366–372).

[48] Marinha compra navio (2008).

[49] Navio Polar Almirante Maximiano (n.d.) Journalistic sources give the amount of 71.5 million reais instead of the 69 registered on the Navy Archives.

[50] The NPo (Polar Vessel) *Almirante Maximiano* have a length of 93.4 m, a beam of 13.4 m, displacing 5,540 tons and reaching speeds of 11.5kt. It had the capacity for 119 people (crew included).

of the PROANTAR. The *Maximiano* not only had a better infrastructure for scientific and logistic activities, but also would be added to the operations of the *Ary Rongel*. Thus, after twenty years of operating with a single ship—following the end of the operations of the University of São Paulo's Oceanographic vessel, *Prof. Besnard*, in January 1988—the PROANTAR once more had two polar-capable vessels. However, the necessary equipment and structures required for the scientific work and the logistical needs of the Navy in Antarctica entailed important works that took until the next year to be finished.

In addition to the new vessel, the PROANTAR incorporated, in 2008, two academic centres specialized in Antarctic studies: the *Instituto Nacional de Ciencia e Tecnologia da Criósfera* (National Institute of Science and Technology of the Cryosphere),[51] based at the Federal University of Rio Grande do Sul, and the *Instituto Nacional de Ciencia e Tecnologia Antártico de Pesquisas Ambientais* (National Institute of Science and Technology for Antarctic Environmental Research), based at the Federal University of Rio de Janeiro. Both institutes arose from a national call for the establishment of scientific institutes of excellence in strategic areas for national scientific and technological development, with the potential to become world-class centres of excellence.[52] While the institutes represented a small portion of the 122 proposals approved nationally, it was a definitive step on the institutionalization and an explicit acknowledgement of the importance of Antarctic science as a field in its own right.

Additionally, the two institutes added a factor of stability to the Brazilian Antarctic science by creating a scientific research agenda that would not depend on the interests of individual researchers, as had been the case up until then. In that regard, the two institutes gave a baseline of stability for the development of long-term research programmes and, as the budget for the Institute was independent from other funding, it offered another layer of financial resources that would improve the overall scientific performance of the PROANTAR.

Therefore, by 2008, the PROANTAR counted with the regular budget of the Navy and the Ministry of Science and Technology, the funding

[51] Actually, the INCT da Criósfera is 'polar' and thus it includes also the Arctic.

[52] Edital No 15/2008—MCT/CNPq/FNDCT/CAPES/FAPEMIG/FAPERJ/FAPESP /INSTITUTOS NACIONAIS DE CIÊNCIA E TECNOLOGIA.

added by the Ministry of Environment, the support of the Petrobras and Oi-Telemar,[53] the extraordinary credit obtained through amendments to the national budget promoted by the Congressional Group, the extraordinary funds given by the Presidency to improve the profile of the PROANTAR and the budget offered to the National Institutes of Science and Technology. These golden years of multiple sources of funding allowed the expansion of the PROANTAR and the renewal in terms of scientific proposals, infrastructure and possibilities. Whether the multiple funding sources gave the PROANTAR increased resources and flexibility in terms of financing its operations, it also made the programme more vulnerable to changes in the economic and political landscape, revealing that, even when heavily funded, the promoters of the programme failed in providing it with a more endurable and stable source of resources. To this must be added that the meetings of the higher political institution of the Brazilian Antarctic policy, the CONANTAR, were suspended in 2008—without resuming the following years—depriving the programme for its main resource of political articulation.[54]

Notwithstanding the deactivation of the CONANTAR, the PROANTAR continued to expand. In August 2009, the CNPq launched a new call for scientific proposals of the PROANTAR.[55] The call presented a qualitative leap in terms of the resources offered, thanks to the support provided by the Congressional Group through special amendments to the national budget, reaching a total of 14 million reais (about USD 7.5 million) for the period 2009–2011—compared with 1.64 million for the 2005–2008 period (about USD 700,000). The call also presented two relevant innovations from previous calls: the prioritizing of partnerships with other South American countries; and the incorporation of the research area of 'Technological, cultural and socio-economic aspects in Antarctica.'

The inclusion of social and cultural aspects was a first in the Brazilian Antarctic programme, up until then centred only on the three main areas of physical sciences, geosciences and life sciences. That shift allowed the creation of the *Laboratório de Estudos Antárticos em Ciências Humanas*

[53] Petrobrás provide oil for the operation of the vessel and the station and Oi-Telemar provided the telecommunication system, all with no cost for the programme.

[54] Ata da 186ª Sessão Ordinária (2014, p. 14).

[55] Edital MCT/CNPq N° 023/2009. Brazil.

(Antarctic Studies Centre on Human Sciences) of the Federal University of Minas Gerais, which would start to develop a wide-scope project of archaeology and anthropology in the Antarctic, under the leadership of Dr Andrés Zarankin.

The same month, the Brazilian Congress approved the creation of the Secretary of the Antarctic Treaty in Buenos Aires,[56] ending the process of ratification of the measure 01/2003 that instituted such a body. Brazil was one of the supporters of the Argentine candidacy to host the Secretary in Buenos Aires since its first proposal in the XVII ATCM of 1992 and was key to obtain the necessary consensus, reached in 2001, that led to its creation during the XXVI ATCM.[57] However, the ratification by the Brazilian congress was delayed by the fact that the installation of the Secretary would require financial contributions from the Brazilian government. Despite the fact that the Secretary was installed de facto in September 2004 in Buenos Aires, ratification by Brazil was necessary to give full thrust to the institution, especially in its financial aspects.

In September, the CNPq called for proposals for trilateral cooperation between Brazil, India and South Africa.[58] Following what was agreed in the meetings of the Ministers of Science and Technology and the corresponding agreements, Antarctic research in oceanography was one of the possible areas to be invested in. Therefore, even if not completely directed to Antarctic science, the another 800,000 reais (about USD 450,000) was added to the possible funds to be available for Antarctic research under cooperative international schemes.[59]

In the meanwhile, the Brazilian infrastructure for its Antarctic activities continued to improve. After many adaptations that included the installation of five laboratories, scientific and oceanographic equipment, navigational and communication systems and a flying deck, in 2009, the polar vessel *Almirante Maximiano* was incorporated to the Antarctic operations.[60] It sailed on 19 October 2009, from Rio de Janeiro, as part

[56] Decreto Legislativo N° 597 (2009).

[57] Desptel 506 para Brasemb Buenos Aires, em 26/07/2007. Arquivo Intradocs. Cited by Ferreira (2009, p. 149).

[58] Edital MCT / CNPq N° 045/2009.

[59] The 800,000 reais were allotted to the whole call, and the maximum for each project was 100,000.

[60] The estimated total cost spent on the NAPOc *Almirante Maximiano* was about 140 million reais (Navio Polar Almirante Maximiano, n.d.).

of the OPERANTAR XXVIII (Brazilian Antarctic expedition). The 2009–2010 expedition experienced several setbacks, including faults in both vessels and the cancellation of the air transport flights by C-130s, due to their transfer to Haiti to assist in humanitarian tasks after the earthquake that devastated the island. Nonetheless, the programme attained its main objectives, including the scientific projects and important adaptations to the EACF. During the winter, a covered garage was added to the station, new dorms were installed, the labs were modernized, and the generators were completely replaced with newer equipment, which were expected to reduce significantly the environmental impact, especially in the reduction of noise.[61]

But while investment and activities in Antarctica were increased, there was a sense that such changes were unplanned, and the programme needed a more systematic approach to the planning of its future. In view of the extension and diversification of Brazilian Antarctic science and the intensification of its political involvement in Antarctic forums, on 26 August 2010, the CIRM established the 'working group PROANTAR,' under the coordination of the Secretary of the CIRM.[62]

The working group was tasked to elaborate a strategic plan for the PROANTAR to be presented to the CIRM, the subcommittee for the PROANTAR and the CONANTAR. The idea represented a change in the reactive style that, up until then, characterized the Brazilian involvement in Antarctica. Taking advantage of the favourable political context, the Brazilian Antarctic community decided to think ahead and try to establish a proactive agenda into the middle and long terms. The working group was established with the inclusion of representatives of the Ministry of Education and Culture, the Ministry of Science and Technology, the Ministry of Environment, the Ministry of Foreign Affairs, the Ministry of Tourism, the CNPq, the CIRM Secretariat and the prominent Antarctic scientist Professor Rocha-Campos, as representative of the Antarctic scientific community.[63]

In October 2010, as the two polar vessels were under-going substantial maintenance work, the Brazilian Navy sent its Submarine Rescue Ship

[61] Ata da 173ª Sessão Ordinária (2009, p. 12), Ata da 174ª Sessão Ordinária (2010, p. 16), and Ata da 176ª Sessão Ordinária (2010, p. 17).

[62] Resolução N° 2 (2010) and Ata da 175ª Sessão Ordinária (2010, pp. 5–7).

[63] Ata da 176ª Sessão Ordinária (2010, p. 16).

Felinto Perry to the Antarctic to refuel and resupply the EACF.[64] Despite the ship not being specifically equipped for polar navigation, the operations were successful, and the ship remained in the region until the arrival of the *Ary Rongel* and the *Almirante Maximiano*.

In terms of the domestic politics, the trajectory of 'Lula' da Silva as President favoured the continuity of the party of government, but the political and economic context had changed considerably. On 1 January 2011, the Workers' Party candidate, Dilma Rousseff, acceded to the Presidency of Brazil.[65] The new President had to confront matters that had accumulated during the previous term and that were a consequence of the expansive policies of 'Lula' da Silva's government. During 2009 and 2010, Brazil had overcome the impacts of the world economic crisis thanks to the social policies and the expansion of its internal market. However, by 2011, inflation and public spending threatened the stability of the economy. In order to contain those variables, in February, Dilma Rousseff announced a big cut in public spending that included all Ministries and was extended to Parliamentary amendments to the budget. As the reduction of funds would affect all governmental sectors—the only exceptions were the social programmes and the *Programa de Aceleração do Crecimento* (Growth Acceleration Programme)—the PROANTAR was excluded from many of its sources of funding. After some years of bonanza, the golden years for the PROANTAR were coming to an end.

Despite some disruptions caused by the change in the steering of the government—such as the cancellation of the Antarctic flights of the Brazilian Air Force—the OPERANTAR XXIX was successful, with 18 scientific projects developed by about 220 researchers. Additionally, the National Institute of Science and Technology for Antarctic Environmental Research in collaboration with the Ministry of Environment were advancing in their studies aimed at establishing an environmental management system for the PROANTAR that would observe the regulations established by the Protocol of Environmental Protection in order to certify it through the environmental management international standard,

[64] Ibid., pp. 15–16.

[65] Dilma Rousseff was Chief of Staff of the Presidency under 'Lula' da Silva and was his nominee for the Presidency on the Workers' Party. 'Lula' da Silva was prevented from acceding to the Presidency for a third term due to the constitutional limit to two consecutive mandates.

ISO 14001.[66] The works of the PROANTAR working group continued, with workshops during Mai and July.

Furthermore, regionally, the approach of Dilma Rousseff to regional neighbours, Argentina and Chile, created prospects for deepening bilateral collaboration in Antarctic issues.[67] While regional collaboration was already important—thirty-three regional partnerships, including fifteen with Argentina and ten with Chile, on the last call for the PROANTAR—new institutionalized possibilities of collaboration were expected. On 31 January 2011 the Presidents of Argentina and Brazil, Cristina Fernández de Kirchner and Dilma Rousseff, signed a joint declaration that included the commitment of both countries to maintain and deepen bilateral cooperation with the intention of consolidating a joint programme and to coordinate their political positions within the ATCMs and the CCAMLR.[68] Also, in a bilateral meeting between the PROANTAR and the Chilean Antarctic Institute during the COMNAP, Chile offered the use of their station, Teniente Luis Carvajal Villaroel, for the development of joint projects in southern latitudes and their collaboration in future Brazilian expeditions to the interior of the continent.

This offer included the already committed collaboration with the expedition planned by the glaciologist Jefferson Simões, of the Federal University of Rio Grande do Sul, director of the National Institute of Science and Technology of the Cryosphere. Simões' project, 'Cryosphere expedition,' in collaboration with the Chilean Antarctic Institute, took place between December 2011 and January 2012. The project, that cost 1.2 million reais (about USD 670,000), included the installation of an automatic meteorological and atmospheric module in the interior of the continent, at 84° south latitude, and the collection of ice-cores and geological samples during the expedition.[69] Being the first such project developed by Brazil, the expedition inaugurated a new era of exploration of the Antarctic inner mainland and the expansion of the infrastructure to southern latitudes (see Fig. 5.1), notwithstanding future financial strains putting a halt to that expansion.

[66] Ata da 178ª Sessão Ordinária (2011, pp. 23–24).

[67] Ibid., p. 21.

[68] Declaração Conjunta (2011).

[69] Ata da 178ª Sessão Ordinária (2011, pp. 26–27).

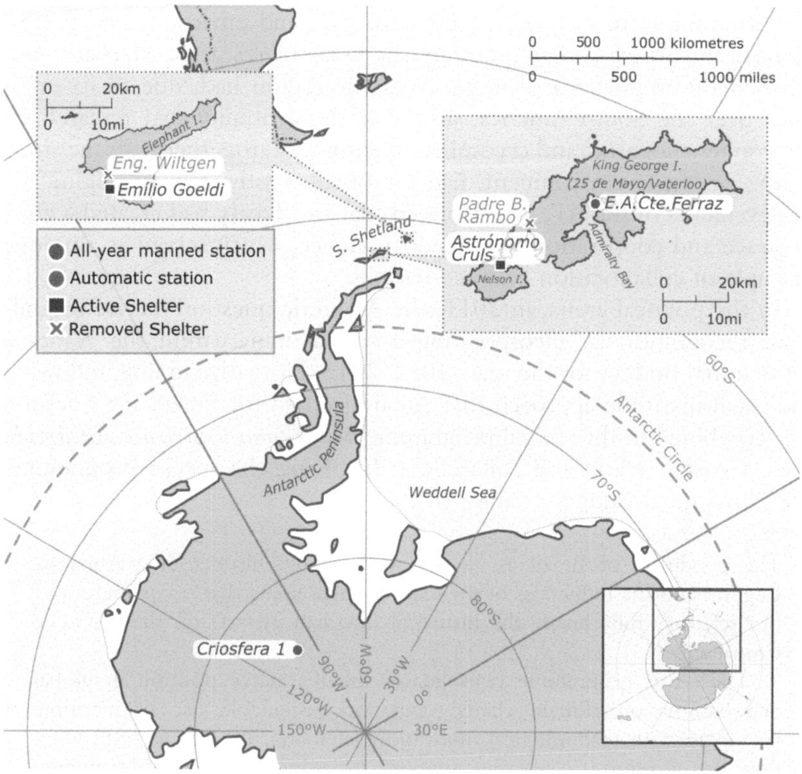

Fig. 5.1 Map showing Brazilian infrastructure in Antarctica (*Source* The author)

Between 20 June and 1 July 2011, the ATCM XXXIV took place in Buenos Aires, celebrating the 50th anniversary of the came into force of the Antarctic Treaty (AT). In the opening speech, the head of the Brazilian delegation, Ambassador Luiz Alberto Figueiredo Machado, praised the treaty, acknowledging the important milestone that was being celebrated and taking advantage of the occasion to announce the upcoming 30th anniversary of the PROANTAR.[70] He gave a succinct account of the evolution of Brazil's involvement in Antarctica and within the Antarctic regime, and highlighted the recent investment that allowed

[70] Statement by Ambassador (2011).

the programme to improve its infrastructure and environmental performance, and to expand significantly the areas of research. Machado also stressed the importance that the Antarctic region had, due to its influence over the South America, justifying the centrality that research on atmospheric, oceanic and cryosphere phenomena, together with the study of the Antarctic environment, had for Brazil. Lastly, acknowledging the achievements of the AT, he celebrated that the Treaty had created a zone of peace and cooperation dedicated to science, which served as a unique example of collaboration between nations.

In the political arena, in 2011, the Antarctic question received significant recognition by incorporating a special item within the National Pluriannual Budget for the years 2012–2015. Notwithstanding, following the usual maritime approach that Antarctica had for Brazil, the question was combined in the special programme *Mar, Zona Costeira e Antártida* (Sea, Coastal region and Antarctica). Justifying the special programme, the Pluriannual Budget established that:

> The condition of Brazil as an Atlantic country, close to the Antarctic Region, and the influences of the natural phenomena that occur there over our territory fully justify the historical Brazilian interest on the Antarctic Continent. (...)
>
> Thus, this programme consolidates Brazil's active position in global negotiations on climate change and the sustainable use of maritime resources, with technological innovations enabling new chains of production, in addition to offering support for conducting multidisciplinary research on the Antarctic Region and its adjacent area.[71]

The inclusion of the specific item responded to the political effort of the Ministries of Science and Technology, Environment and Defence, with the objective of emphasizing the strategic and political significance of the issues included within the competences of the CIRM and provide them with a wider horizon of financial stability. The focus was on scientific research, sustainable development, environmental preservation, oceanographic and climatological monitoring, human resources formation and maritime mentality.[72]

[71] Plano Mais Brasil (2011, pp. 271–272, own translation).

[72] Ata da 178ª Sessão Ordinária (2011, pp. 31–32).

Complementing that long-termed vision, in November 2011, the CIRM approved the report of the working group PROANTAR, which included the strategic plan for the years 2012–2022.[73] The report justified that:

> On the eve of its 30th Anniversary, the Brazilian Antarctic Programme (PROANTAR) is at the moment of rethinking its future and orient its priorities according to the needs of the scientific community, the Brazilian society and the national interest.[74]

The report acknowledged the relevance that the political agreement of freezing the territorial status quo of the continent had in making possible for Brazil to undertake activities in Antarctica and it considered that, at least in the following 10 years, no significant change would be expected in the international regime. Nonetheless, it also considered that:

> ...the growing Brazilian interests in the South Atlantic and, in particular, in the Antarctic continent, maintain the historical, current and future strategic concerns, which transcend the exclusively scientific interests, without, however, be exempted from them in any form. In this sense, the scientific and technological knowledge developed and accumulated will be decisive for the country's progress towards the exploration of the scientific frontiers of the interior of the Continent and the Southern Ocean, leveraging the potential enclosed in the marine and continental natural resources of Antarctica.[75]

Analysing the strengths and weaknesses of the Brazilian programme, the report considered as the main favourable factors that of: (1) being a State policy, having a great visibility in the political spectrum; (2) being considered of strategic importance; and (3) having developed relevant capabilities after thirty years of uninterrupted presence in Antarctica, particularly in scientific development.[76] In respect to the unfavourable factors, the report stressed: (1) the small number of specialists and the high level of turnover; (2) the lack of a centralized management system;

[73] Relatório do Grupo de Trabalho (2011).

[74] Ibid., p. 3, own translation.

[75] Ibid., p. 20, own translation.

[76] Ibid., p. 50.

(3) the reduced visibility for Brazilian society; (4) the instability and deficiency of funding; and (5) the deficiency of infrastructure and the logistical dependency on foreign countries, in particular with Chile. These characteristics, inherent to the PROANTAR, were coupled with the negative impacts of climate change, the increasing number of tourists in Antarctica and the lack of regulation of the Protocol of Environmental Protection. On the other hand, opportunities were offered to Brazil by its relative closeness to the Antarctic region, the potential for international cooperation and the increasing environmental conscience that aroused globally.

As a guideline for the future, the report established the vision of 'becoming a programme of international reference and leadership within the Antarctic Treaty System, due to the quality of its scientific research, giving emphasis to those aspects that have influence over Brazil.'[77] As had happened in the past, the proposal emphasized the need for Antarctic research to be linked to direct benefits in the Brazilian territory, including meteorological conditions, the prediction of potential impacts from climate change and changes in the Antarctic marine biodiversity, the development of industrial products through bioprospecting, among others. It is necessary to note that no consideration was given to aspects related to humanities and social sciences, with the only exception of a single mention of the intention to implement a communication programme oriented towards promoting awareness and educational content to the public on what regarded as the importance of the Antarctic environment, with emphasis on its potential effects for South America and Brazil.

Following that vision, the report proposed to follow three main strategic objectives: (1) prioritize support for qualified, multidisciplinary and long-term scientific demands; (2) expand the geographical scope of PROANTAR's scientific activities in the Antarctic region; and (3) promote scientific partnerships, exchanges and cooperation agreements for Southern polar research, giving priority to South American countries.[78] Furthermore, it was expected that the advancement of the PROANTAR in the next ten years could ensure: (1) Brazil's participation in the ATS; (2) the undertaking of multidisciplinary Antarctic research of

[77] Ibid., p. 21, own translation.

[78] Ibid., pp. 23–24.

international quality and related to Brazil's interests; (3) the development of technological innovations applicable to Antarctica; (4) the protection and conservation of the Antarctic environment; (5) the formation of human resources in Antarctic fields and the support for Antarctic research centres; (6) the availability of logistic capabilities that allow operations under safety conditions; (7) the expansion and consolidation of international cooperation in Antarctic matters; (8) the acknowledgement of PROANTAR achievements by Brazilian society; (9) the effective control and monitoring of the Brazilian flow of tourism in Antarctica; and (10) the integration of all actors involved in the programme.[79]

Besides these general orientations, the report also advanced some more specific suggestions. Among many recommendations, the document suggested: (1) coordination between the Ministries and the participation of their delegates in all relevant international instances; (2) the call to issue-specific and open scientific proposals, with multiple funding and long-term planning; (3) the call to multinational projects in areas of common interests; (4) widening the scheme of partnerships; (5) the use of alternative energy sources and a more efficient waste disposal; (6) the creation and consolidation of centres specialized in Antarctic research; (7) the acquisition of a third polar vessel with capacity of navigation through thicker ice; (8) the substitution of the EACF with a new and more efficient station by 2022; (9)the development of capabilities for air operation in the inner continent; (10) the promotion of public awareness through mass media outlets and the inclusion of content in the educational curriculum; (11) the creation of a legal framework for Antarctic tourism for companies operating from Brazil; and (12) a better integration through a centralized information system and a deeper involvement of all ministries concerned, including financially.[80]

In conclusion, the report also recommended that the 1987 Brazilian Antarctic Policy be reviewed to assess the necessity of its reformulation; the revision of the current Brazilian Antarctic programme in the light of the advancements of the ATS and the strategic plan; and the assessment of the National Policy for Antarctic Science.[81] Despite the financial difficulties, the PROANTAR had finally adopted a proactive and strategic

[79] Ibid., p. 25.

[80] Ibid., pp. 26–40.

[81] Ibid., p. 49.

approach and new prospects seemed to open, as the 30th anniversary of Brazil's presence on the continent approached. However, the events of the OPERANTAR XXX proved those expectations were hasty.

5.3 Rebuilding from the Ashes: The Tragedy of the OPERANTAR XXX and the Contradictions of the Brazilian Antarctic Programme

The OPERANTAR XXX, that took place on the 2011–2012 season, coincided with the 30th anniversary of the launching of the PROANTAR and was characterized by a series of unprecedented incidents in Brazil's Antarctic record. The first setback occurred early in the season, on 29 October 2011, when a raft used to transport the diesel from the *Maximiano* to the station sank with some 10,000 litres of oil. The following day, the *Maximiano* located the raft and placed floating barriers to contain any potential spill.[82] While the accident did not result on any loss of life or environmental impact, it was the first of such incidents for Brazil and would pale in comparison with what was about to happen.

As provided in the strategic plan, looking for a better environmental performance of its Antarctic operations, the PROANTAR decided to experiment with alternative energy sources. Between December 2011 and January 2012, an experimental ethanol generator was installed in the EACF to complement the recently replaced diesel generators. The new equipment would undertake a series of technical tests to assess the viability of its use in Antarctica. It was provided by the *Vale Soluções em Energia*, a company belonging to the Vale Mining Company and the Brazilian Development Bank, in partnership with the PROANTAR.[83] The project was expected to increase Brazil's stand in front of the Antarctic community through the development of technologies that would made the use of biofuels in the continent possible. However, the project was interrupted by unexpected events.

With the situation of the raft contained, the advancements of the expedition were going according to the plan until the 25 February 2012, when the tragedy struck the Brazilian Antarctic programme. At dawn, a fire in the module of generators spread to the rest of the station. The night

[82] Ata da 180ª Sessão Ordinária (2012, pp. 18–19).

[83] Ata da 179ª Sessão Ordinária (2011, p. 16).

before, one of the Navy officials was doing a routine transfer of fuel from the main tanks to the generator room. Allegedly, the official decided to attend a party that was being given at the EACF instead of supervising the whole process, resulting in an overspill of oil in the generator room, which started the fire.[84] The incident destroyed about 70% of the EACF and took the life of two officers who fought to control the fire to ensure the safe evacuation of the station. Only three isolated modules survived the fire: the Chemical Lab; the Very Low Frequency Lab; and the Meteorology Lab. Despite the fire being attributed to a single human error, various questions were raised about the conditions that allowed such a tragedy,[85] while the loss of life left an indelible mark in Brazil's Antarctic programme.

The fire at the EACF was a heavy blow for the PROANTAR. Brazil's home in Antarctica was ruined, samples from some scientific projects were destroyed and two lives were lost. However, it was not the end of the Brazilian programme and the OPERANTAR XXX. With the assistance of Argentina, Chile, Poland, UK and Uruguay, Brazil moved the personnel and cargo from Antarctica and initiated the removal of the station's remains to contain any possible contamination to the environment.

But the fire was not the only problem of the PROANTAR. The *Ary Rongel* was kept docked in Punta Arenas in January and February due to mechanical problems, and the *Maximiano* had to remain in the proximity of the sunken oil raft in order to prevent any leakage. In the last days of February, the submarine support vessel *Gulmar Atlantis,* lent by Petrobras, started preparations to refloat the oil raft, which was finally achieved on 3 March.[86] However, this was not the end of Brazilian problems in Antarctica.

[84] The investigation was undertaken under the Military Justice and remained classified due to 'security and strategic concerns.' The accounts correspond to the information that appeared in the press. See, for example: Base da Antártida estava em festa (2012) and Tribunal Militar condena sargento (2016).

[85] Some raised the possibility that the fire alarm system had been disabled intentionally to avoid interrupting the party in case of false alarm. Additionally, it was considered the possibility of a connection between the experimental ethanol generator and the fire, a possibility that was discarded by the Navy. Whatsoever, the project of the biodiesel generator for Antarctica was abandoned after the fire, suggesting that there may indeed have been a connection. In any case, the lack of access to the investigation makes it impossible to assess the conditions that caused the accident.

[86] Ata da 180ª Sessão Ordinária (2012, pp. 18–19).

On 7 April 2012, the tourist motorboat *Mar sem Fim*, sailing under the Brazilian flag and captained by a Brazilian businessman, sank in Maxwell Bay, in front of the Chilean station Presidente Eduardo Frei Montalva.[87] The ship had delayed its journey through Drake Passage due to the meteorological conditions and was cracked by the ice brought to the bay by the strong winds. To aggravate the situation, the shipwreck showed signs of fuel leakage and the voyage had not been carried out with the necessary notifications to the PROANTAR and the authorization of the Ministry of Environment. Added to that, the next day, the *Ary Rongel* was surrounded by thick ice in the Bransfield Strait (*Mar de la Flota*), causing a request for assistance that was responded to by the British Antarctic research vessel *Ernest Shackleton*. Thanks to the aid of the British vessel, the *Ary Rongel* was able to navigate the thick ice and reach the open sea.

Despite all these setbacks, the OPERANTAR XXX managed to complete many of the planned scientific activities. One of its most relevant achievements was the success of the expedition to the interior of the continent lead by Jefferson Simões and supported by the Chilean Antarctic Institute. The expedition took important samples and installed the automatic module 'Criósfera 1.' The module was the first infrastructure installed by Brazil in the interior of the continent and represented an important step in widening the geographical scope of the Brazilian Antarctic programme, as it could be seen in Fig. 5.1.

To keep a formal presence on the continent, the Brazilian Navy decided to leave four officers who remained housed in the Chilean station Frei during the winter season.[88] As a reaction to the demise of the EACF, on 7 March 2012, the Brazilian government issued a resolution ordering the allocation of an extraordinary fund of 40 million reais for the recovery and reconstruction of the station.[89] This was complemented by the instruction to constitute an inter-ministerial working group to establish the specifications of the new EACF, to:

> ...observe, in particular, the scientific needs for Antarctica, the environmental standards set in the Antarctic Treaty and its Protocol of

[87] Ibid., pp. 19–21.

[88] Informe Final (2012, p. 7).

[89] Medida Provisória N° 560 (2012).

> Environmental Protection (Madrid Protocol), and the logistic specificities intrinsic to the Antarctic environment, stressing its functionality and the safety of personnel and installations.[90]

As the fire raised the profile of the PROANTAR in public opinion, other political developments institutionalized a more important role for Antarctica within the strategic concerns of the country, albeit in a minor discursive way. By the middle of 2012, a revision of the National Defence Policy of Brazil was presented to Congress for its consideration and approval. Different to the previous version, the new policy included Antarctica within the 'strategic surroundings' of the country, jointly with the South Atlantic and the 'bordering' [*sic*] countries of Africa.[91] Jointly, the Government presented the Defence White Paper,[92] in which the Antarctic region was considered within the '...strategic maritime areas of greater priority and importance to Brazil....'[93] The document stressed once more the influences that the Antarctic region exerted over the Brazilian coast and its maritime living resources, justifying that '...due to the unparalleled interest the region has for the country, the Brazilian Antarctic programme (PROANTAR) was approved in January 1982.'[94] It is relevant to highlight that, once again, the White Paper grouped the issues of sea, Antarctica and outer space in a single item within the twenty-first-century Strategic Environment, repeating the grouping that had characterized the country's approach.

At the end of September 2012, the XXIII RAPAL took place in Rio de Janeiro. In the opening address, the coordinator of the CIRM, Commander Julio Soares de Moura Neto, took the opportunity to thank the expression of solidarity and offers of cooperation from the fellow programmes, considering the timing propitious to the '...community of

[90] Portaria Interministerial N° 1199 (2012, own translation).

[91] Política Nacional de Defesa (2012, Item 4.1).

An analysis of such inclusion and its relevance for Brazil can be found in: Mattos (2014) and Abdenur and Neto (2014).

[92] In the presentation of the document by the President, the White Defence Paper is regarded as a tool for the society: '...as a whole, to stay abreast of military affairs while conveying the potential and needs of our Defense to the public debate' (Defense White Paper, Message from the President of the Republic, 2012).

[93] Defense White Paper (2012, p. 38).

[94] Ibid., p. 48.

efforts of the Latin American nations, with a view to finding solutions to the issues that affect the Antarctic.'[95] Matching the tone, the Minister of Defence, Ambassador Celso Amorim, gave thanks for the cooperation received and ensured the Brazilian commitment in Antarctica. During the event, Brazil highlighted the common front and discourse presented by the Latin-American countries and called for further collaboration exceeding merely operational aspects.[96]

The occasion was used not only to establish the Antarctic partnership between Latin-American countries, but also to make explicit the relationship between Brazil and Antarctica. During the meeting of the Commission of Logistic and Operative Issues, Brazil insisted on establishing the bond between Antarctica and Brazil through the environment, with the justification that: '…as a tropical country, there is a need to explain what Brazil does in Antarctica: the climatic dynamics shows the direct influence that exists between Antarctica and Brazil.'[97]

The fire at the station did not curtail the scientific work of the OPERANTAR XXXI, which was considered the biggest and more complex operation to that date.[98] The operation gave support to twenty-one projects that involved some 200 scientists, in addition to the complex works involved in the removal of the remains of EACF and the installation of emergency modules to continue the year-round operations. Due to the numerous tasks involved, the Brazilian Navy made use of a number of vessels. The *Maximiano* was exclusively dedicated to research activities, the *Ary Rongel* to provide assistance to logistical and scientific activities, and the *Felinto Perry* was in charge of the transport of materials and personnel and the support of the operations of refloating the *Mar sem Fim*. Another two vessels, the merchant ship *Germania*, chartered by the CIRM Secretariat, and the Argentine Cargo Ship *Bahía San Blas*, offered in cooperation by the Argentine Navy, were added to support the logistics of the operation. Argentina also offered its installations at the Camara base, on Livingston Island, where seven Brazilian remained until the beginning of December.

[95] Informe Final (2012, p. 25, own translation).

[96] Ibid., pp. 7–8.

[97] Ibid., p. 47, own translation.

[98] Ibid., p. 17; and Ata da 183ª Sessão Ordinária (2013, p. 7).

The dismantling of the EACF required detailed planning to comply with the environmental standards set by the Protocol of Environmental Protection. The Ministry of Environment monitored all the activities and ensured that the impact of the works was minimal. The removal of some 900 tons of waste was completed on 17 March 2013, when the *Germania* departed from Admiralty Bay to transport the cargo to Rio de Janeiro, where was finally disposed of. By that time, 45 emergency modules were already installed at the heliport. The emergency modules and the three remaining modules of the EACF restored the facilities for the Brazilian year-round activities to a total area of about 1000 square meters, ensuring the continuity of the Brazilian presence on the continent and the scientific works developed in the area.

Although the fire at the EACF was a heavy blow, and the process of removal, the installation of emergency modules and the planning of a new station were demanding significant resources, the scientific community did not interrupt the process of planning the future of Brazilian Antarctic science. In 2013, the CONAPA took the strategic plan developed by the CIRM as a basis for the elaboration of a roadmap for Antarctic Science: 'Antarctic Science for Brazil: Action Plan 2013–2022.' The action plan defined as its vision:

> To become an internationally recognized nation thanks to its top scientific performance in the Antarctic and Southern Ocean region, implementing thematic programs in a sustainable manner, investigating the environmental processes and current, past and future relationships between the South American continent and the polar regions.[99]

Complementing this vision, the central mission of the roadmap was defined following an instrumental political role for Brazilian Antarctic science, with the aim to '…help to guarantee the permanence of Brazil as a consultative member of the Antarctic Treaty.'[100]

The plan oriented future Antarctic research in five programmes, following the SCAR permanent science groups and aimed at exploring the connections between the Antarctic continent and South America: (1) the role of the cryosphere in the Earth system and its interactions with South America; (2) the effects of climate change on the bio-complexity

[99] Simões et al. (2013, p. 3, own translation).

[100] Ibid.

of Antarctic ecosystems and their connection with South America; (3) climate change and the Southern Ocean; (4) geodynamics and geological history of Antarctica and its relations with South America; and (5) dynamics of the upper atmosphere in Antarctica, interactions with geo-space and connections with South America.

The focus on the connexions between the Antarctic and South America took the emphasis given to those aspects on the CIRM strategic plan and sought to provide some direct connection between the scientific activities in the polar regions and the Brazilian mainland, in order to justify the investment made in Antarctica through the directs benefits that the Brazilian society could accrue directly or indirectly. However, that connection was made through physical and biological environmental links, setting aside all humanities and social sciences and depriving the Antarctic programme from a field that was obtaining increasing attention internationally.

Notwithstanding the loss that the fire at the EACF represented in material, human, scientific and international prestige terms, the accident presented a unique opportunity for renovation. It brought the conditions and limitations of the programme—especially concerning the safety conditions of operations and infrastructure—to public attention and pushed the government to commit significant resources to the reconstruction of the station. This accelerated the planned replacement of the station with a more modern installation, that was foreseen in the 2011 strategic plan, but which had no guaranteed funding. At the end of 2012, a workshop and a seminar helped to define the technical aspects that would define the call for proposals for the new EACF. Elaborating on these bases, on 28 January 2013, the *Instituto de Arquitetos do Brasil* (Institute of Brazilian Architects) launched a competition for preliminary projects for the station. Out of the seventy-four projects presented, the winner that presented by architect Fábio Henrique Faria, from the architect studio of Curitiba, *Estúdio 41 Arquitectura SS Ltda*, which immediately started work on the executive project to be included in the bidding.

The accident at the EACF also cemented the partnership between Brazil and its main regional partners, Argentina and Chile. Conversations for furthering cooperation continued with Argentina, and Brazilian members spent another season at the Argentine station Camara. More significantly, on 26 January 2013, Brazil and Chile signed an agreement on Antarctic cooperation, the first of such accords signed by Brazil. Up until then, Antarctic cooperation was limited to single mentions in international declarations and cooperative agendas. In contrast, the agreement signed with Chile specifically established a wide framework of Antarctic cooperation, including: the realization of joint activities to optimize the use of resources and diminish the environmental impact; cooperation in technical and scientific activities; interchange of information in fields of common interest, such as technological development and Antarctic environment; the common use of logistical capabilities and the organization of joint expeditions; and joint calls for scientific projects, among others.[101]

Brazil's careful removal of the fire debris and the environmental restoration work paid off, gaining international recognition and approval. In May 2013, during the XXXVI ATCM in Brussels, Belgium, Brazil was praised by its efforts removing the remains of the EACF, the installation of the emergency modules and the attention given to the environment.[102] Although the accident had interrupted an impeccable track record in Antarctica, the Brazilian reaction allowed it to rapidly regain international prestige and display itself as strongly committed to the Antarctic environment.

However, the Brazilian programme was not limited to the reconstruction of the EACF and the budget allocation for scientific projects from 2009 were exhausted. Thus, in September 2013, the National Research Council, the Ministry of Science, Technology and Innovation and the National Fund for Scientific and Technological Development launched a call for proposals with a total budget of 13.8 million reais (about USD 6 million).[103]

Different to the previous call, the research areas to be financed were distinguished between priority and emerging areas. Between the first were defined fields within the natural sciences, always stressing their relations with South America, including: the interactions between the ice and the atmosphere; the effects of climate change on the Antarctic ecosystem; the trends and vulnerability of the Southern Ocean; the Antarctic geology and the upper atmosphere dynamics. Between the emerging areas were those disciplines linked to technological developments and social sciences, including: the prospecting of extreme organisms; the microbiology of

[101] Acordo de Cooperação Antártica (2013).

[102] Ata da 184ª Sessão Ordinária (2013, pp. 18–19).

[103] Chamada/Ação Transversal n°64 (2013).

anthropogenic origin in Antarctica; human medicine and psychology in Antarctica; architecture, civil constructions and new materials; and anthropology, archaeology, sociology of science and public policy.

The call was followed by a new statute for the CONAPA, approved by the Ordinance N° 31 of the Ministry of Science, Technology and Innovation on 13 January 2014. This updated the composition of the committee, including the representative of the Secretary of Aquiculture and Fishing—introduced de facto in 2009—and complemented some of the previously defined regulation. Despite that the new norm did not alter significantly the operation of the committee, the appointment as representative of the National Research Council to the '...Director of Agricultural, Biological and Health Sciences, or his representative,' is of some attention, especially because that determination excluded scientific disciplines of traditional Antarctic relevance, such as the Physical Sciences and Geosciences, from leading the CNPq representation within the body.

Nevertheless, these disciplines were not left without representation, as the statute established that scientists from fields '...aligned with the areas of knowledge defined by SCAR, for example, Life Sciences, Geosciences and Physical Sciences' should be part of the seven representatives of the scientific community.[104] Yet, whether the new statute did not exclude the humanities and social sciences, the emphasis given to the scientists from the natural sciences within the committee seemed to consolidate the predominance of those fields of Antarctic science within its main scientific institutional representation, despite the recognition as 'emerging field' in the recent call of the PROANTAR.

Added to the renovation of the CONAPA, after some years of inactivity, the operation of the CONANTAR was resumed. The reactivation of the political instance of debate of the Brazilian Antarctic programme was necessary in the light of the challenges that it had confronted after the fire, and due to the hosting of the XXXVII ATCM in Brasília, between 28 April and 7 May.

The ATCM was an opportunity for Brazil to renew its Antarctic commitment internationally and to display the recent advancements. Therefore, during the meeting, the country was able to inform the other parties of the advancements of its scientific programme, the removal of the debris of the former EACF and the environmental remedies which

[104] Portaría N° 31 (2014).

had been implemented, the project for the new station, the outcomes of the research on pollutants in Admiralty Bay and the presentation of the action plan for Brazil's Antarctic science. That way, Brazil not only showed the continuity of its scientific work and the concern with the preservation of the Antarctic environment, but also its long-term view about the country's presence and activity in the region.

Following Brazil's intention to have a more active role in the Antarctic regime, the OPERANTAR XXXIII visited a few foreign Antarctic stations with a delegation composed of representatives from the Navy, the Ministry of Foreign Affairs, the Ministry of Science, Technology and Innovation, the Ministry of Environment and the Head of the Argentine Antarctic programme.[105] The visits were a decision taken during the RAPAL 2012 and included: the Argentine stations Carlini and Camara; the Chinese station Great Wall; the South Korean station King Sejong; the Russian Bellingshausen station; the Chilean station General Bernardo O'Higgins; and the German meteorological monitoring station. In counterpart, the Brazilian emergency modules received an international inspection from the Czech Republic, Spain and the UK, receiving a favourable appraisal of the works developed there.[106] However, the operation was also marked by an accident, this time without casualties. In November 2014, a Hercules C-130 of the Brazilian Air Force suffered an accident when landing in Antarctica. This rendered the plane inoperative and had to be abandoned, until it was possible for it to be dismantled and transported back to Brazil, between November 2016 and January 2017.[107]

But the Hercules was not the only obstacle of the PROANTAR. With the executive project of the EACF finished, the initial bidding for the construction of the station was limited to national companies—or foreign companies with national partners—with a total valuation for the construction of 145 million reais (about USD 60 million). However, the estimated cost and the technical complexities involved in the project seem to have determined that, by the closure of the bidding, in February 2014, the call ended with no proposals. The outcome of the call was a serious setback to the original plan as it delayed the whole process.

[105] Ata da 188ª Sessão Ordinária (2014, pp. 15–16).

[106] Ata da 189ª Sessão Ordinária (2015, p. 18).

[107] Ata da 195ª Sessão Ordinária (2017, Item 6.2, lines 516–537).

To avoid unnecessary further delay, the government studied the possible motives of the previous failure and, in July 2014, the Navy launched a new call. This time, foreign companies would be able to present proposals without the need of any national partnership and the value of the project was updated to up to USD 110.5 million, representing almost double the amount proposed of the first call. In May 2015, after several delays due to bureaucratic deficiencies in the proposals, the Chinese state company China Electronics Import and Export Corporation (CEIEC) won the bidding with an offer for nearly USD 100 million. After some delays, the contract was awarded to the Chinese company on 31 August 2015.[108]

In the meantime, the Brazilian Government was entering a political crisis of enormous proportions. After four consecutive terms and more than twelve years of the Workers' Party being in power, the conservative economically dominant sectors and the right-wing media started to push an institutional coup that could end the era of progressist, social policy-oriented governments and implement a neoliberal reform agenda.[109]

The conspiracy against the President Dilma Rousseff seems to have started as soon she won the re-election, by a small margin, at the end of 2014, but took form on the second half of 2015. Despite being a strong fighter against corruption, even against her own Party, the corruptions scandals that involved the Workers' Party were mobilized by the press and the opposition to inflame the public opinion.[110] As the President's popularity was in free fall, the Vice-President Michel Temer, the President of the Senate, Eduardo Cunha and others from opposition parties proceeded, on 2 December 2015, with a request for impeachment.

Dilma Rousseff was removed from the Presidency on 12 May 2016, after both Chambers of Parliament approved to open an impeachment

[108] The two competitors appealed the decision, delaying the adjudication, but their arguments were rebutted by the adjudication committee.

Documents on the process of bidding could be found in: https://www.marinha.mil.br/secirm/licitacoes#concorrencia-02-2014.

[109] There is abundant literature on the role of mass-media groups in the institutional coup against Dilma Rousseff. See for example: Becker et al. (2016), De Almeida and Lima (2016), Marques et al. (2018), Rodrigues (2018), and Van Dijk (2017).

[110] Accusations of corruption were kept aside from the impeachment trial due to being considered as unsubstantial. Even one of her detractors, the former Federal Prosecutor Rodrigo Janot, declared that he had certainty that Dilma Rousseff was not involved in corruption. See: Janot: 'Não tenho dúvida... (2019).

trial. The trial took until the end of August, when Rousseff was found guilty of administrative misconduct, resulting from the approval of 'illegal' financial operations over the federal budget—the so-called *pedaladas fiscais*—consisting of the approval of extraordinary credits without due sanction of Congress—a sort of overleveraging the budget.

As soon as Rousseff was set aside from office to face the trial, Michel Temer assumed the Presidency and proceeded to replace the Cabinet, starting a neoliberal reform at odds with the political programme that won the re-election in 2014. The rapid changes introduced by Temer, without waiting until the outcome of the trial, signalled the Vice-President was already convinced that Rousseff would be definitively removed.[111] Between his reforms, Temer merged the Ministry of Science and Technology with the Ministry of Communications, despite the great disparity on their respective purposes and the concerns expressed by important sectors of the scientific community, including manifestations against from the *Sociedade Brasileira para o Progresso da Ciência* (Brazilian Society for the Advancement of Sciences) and the *Academia Brasileira de Ciências* (Brazilian Academy of Sciences). That policy was coupled with significative cuts in science and technology public spending, even motivating a letter from 23 Nobel Price laureates, calling for a reconsideration of the planned cuts on Brazilian science.[112]

Thankfully for the PROANTAR, the reconstruction of the EACF depended on the Ministry of Defence, which did not experiment such a dramatic cut on funding. Thus, the advancements in the reconstructions were not interrupted. In contrast, scientific activities were strongly impacted, having to complement the funds granted in 2013 with annual addendums, instead of launching a new call in 2016 as would have corresponded. In fact, cuts in funding have already started before the removal of Rousseff from the Presidency. In April 2016, the situation was so critical that the involvement of the *Maximiano* in the operations were undefined.[113] Furthermore, participation of delegates from the Ministry of Science, Technology and Innovation in the diverse international fora was impeded by the lack of funding.[114]

[111] Se assumir, Temer promote (2016).

[112] Ganhadores do Nobel (2017).

[113] Ata da 192ª Sessão Ordinária (2016, Item 6.1, lines 640–647).

[114] Ata da 193ª Sessão Ordinária (2016, Item 6.4, lines 610–626).

Notwithstanding the shortages in resources, the advances in Antarctic science by the Brazilian programme and the prominent place acquired by the research undertaken in the National Institute of the Cryosphere allowed that, in August 2016, the Professor Jefferson Simões be distinguished by his election as Vice-President of the SCAR. The election of Simões was another acknowledgement of the international Antarctic science community of the quality of the Brazilian programme and its Antarctic scientists. However, the international recognition was not translated in increasing resources. By 2018, the situation was so critical that Simões warned:

> Due to the total lack of financial resources, as of July 2018, the continuity of the scientific work of the Brazilian Antarctic Programme (PROANTAR) is seriously compromised. Since 2013, no specific public call has been issued for national Antarctic research, making the millionaire investment in a new national Antarctic station and vessels counterproductive. Without resources for research, projects, laboratories and, mainly, for maintaining scholarships, the Brazilian Antarctic scientific progress will decline.[115]

The cut in funding made it necessary to extend the three-year budget and make use of the funds of the National Institute of Science and Technology of the Cryosphere, the only such Antarctic institute which funding was renewed in 2014, to cover regular expedition costs and basic emergency maintenance of the module Criosfera 1. The installation of a second module, planned for the 2016/2017 season, had to be suspended. Furthermore, the cuts to the system of science and technology affected the availability of grants and downgraded the professional perspectives of the recently qualified Antarctic specialists, provoking what Simões considered as a 'brain drain.'

In August 2018, the lobby of the scientific community was finally heard and a call of 18 million reais (about USD 4.5 million) for research proposals for the PROANTAR was launched. In this opportunity, funding of proposals was divided in five categories.[116] The first two categories included projects up to 2 million reais per project with a total of 6 million for each category, the first including the cryosphere and the upper atmosphere studies, and the second climate change and the Southern Ocean.

[115] Simões (2018, p.1, own translation).

[116] Chamada N° 21 (2018).

The third category included the studies of the biocomplexity of Antarctic ecosystems, with funding to up to 1 million reais for each project and a total of 3 million for the whole category. The fourth category included projects related to the geodynamic and historical geology of Antarctica to up to 800,000 reais per project and a total of 1.614 million reais for the category. Lastly, the fifth category included the areas of human and social sciences; polar human medicine and biology; and technological innovations, with projects up to 700,000 reais for each project and a total of 1.414 million reais for the category.

The call presented a brief relief to the programme, providing funding for 16 projects for a total of 15 million reais. However, the structure of the call did not satisfy all the proponents, and some areas, such as Humanities and Social Sciences were not granted funding for any single project, even excluding funding for the proposal submitted by the Antarctic Studies Centre on Human Sciences of the Federal University of Minas Gerais. Also, although the call was the greatest made up to that date, in real terms it felt short of other calls, and it paled in comparative terms with the resources committed to the reconstruction of the station.

The election of the right-wing candidate Jair Bolsonaro to the Presidency did not ameliorate the perspectives for public investment in science and technology. Committed to an even deeper neoliberal agenda of reforms and with an explicit and public negationist and anti-scientist discourse, even before his access to power, Bolsonaro identified in the scientific community a rival social group, making it a continuous aim of his attacks. Therefore, his Presidency, which started on 1 January 2019, was marked by a discursive and material attack on science and the academics. In April, Bolsonaro applied a budgetary contingency of 42% to the funds allocated to the Ministry of Science, Technology, Innovation and Communication, leaving the Brazilian scientific and technological system on the verge of collapse. Whether the release of funds at the end of the year alleviated the situation, the main damage was already done, and further cuts were made to the 2020 and 2021 annual budgets in the areas of science, technology and innovation.

The election of Bolsonaro disrupted the economic, educational, social and scientific trajectory of the last decades, which actually had already started to change with Temer's Presidency. More notably, Bolsonaro's foreign policy was marked with the relinquishment of the tradition of autonomy of Itamaraty and the adoption of an international position that followed a reactionary and conservative agenda,[117] an ideological and subservient alliance with the former US President Donald Trump, and the discursive confrontation with important commercial partners, such as China. Luckily for Brazil, that subversion of the traditional lines of Brazil's foreign policy and the downgraded status in the international arena did not reach its position in the Antarctic regime, where there was no disruption, at least to the present time.

Notwithstanding the difficulties imposed by the political situation, the reconstruction of the EACF continued and the projects advanced with the already allotted funds. Despite the contract being granted to a Chinese company, the construction of the new EACF included an important process of interchange and transfer of technology, with the involvement of local engineers in the development of the modules and the installation of station's systems.[118] By 2019, the main buildings of the new EACF were in place and all systems of the station tested. As the project did not included the equipment of the laboratories, the Ministry of Science, Technology, Innovation and Technology allocated 2 million reais (about USD 500,000) for its acquisition. Finally, in January 2020, the new EACF was officially inaugurated in the presence of the Vice-President Milton Mouron and numerous delegations. The new station displayed an innovative design, cutting-edge technology and advanced safety and environmental systems. The installations covered an area of 4,500 m^2 with a capacity to accommodate sixty-four people in summer and thirty-five in winter, with thirty-two dorms, fourteen inner laboratories, five outer laboratories, a fully equipped surgical centre and a library and a common room.

As well as the inauguration of the EACF, in June 2020, bidding for the construction of a new polar ship was launched. The proposal amounted for 750 million reais (about USD 130 million) for the acquisition of a new polar vessel to replace the aged *Ary Rongel*. The ship will be constructed in Brazil, have capacity to navigate through ice blocks—it would not be classified as an ice-breaker—and have a modern system of navigation, being capable of providing an extended range with a reduced

[117] In 2020 Brazil aligned with Saudi Arabia, Bahrain, Qatar, Pakistan, Egypt, Afghanistan and Sudan in the UN Human Rights Council, against the inclusion of paragraphs acknowledging basic women rights in a declaration proposed by Mexico against discrimination against women.

[118] A Experiência da Engenharia (2018).

crew. Despite the acquisition of such vessel will not give Brazil any extension of its operational capacity in terms of geographical reach—due to its limitations to traverse thick ice waters—it will improve the conditions of work and safety of operations.

On the institutional side, in April 2019, an unintended consequence of the Presidential decree 9,759, aimed at simplifying the administrative structure of the State, ended up dissolving the CONANTAR and the CONAPA, the two main institutional political consultative bodies of the PROANTAR.[119] Therefore, the Presidential decision deprived the PROANTAR of two main institutional settings, without interrupting the advancement on the activities of the programme. As a response to the situation, the CIRM assumed the responsibilities of the CONANTAR and established, in July, the creation of the 'technical working group POLANTAR,' in charge of reviewing the 1987 Brazilian Antarctic Policy in order to propose its modification and modernization, taking into consideration the evolution of the Antarctic regime.[120] At last, after some lobbying, the scientific community obtained a favourable view of the reorganization and reactivation of the CONAPA, ensuring the representation of the scientific community through its key institutional interlocutor within the programme.

Despite this institutional embarrassment, some advancements were registered throughout 2020. The transition from the emergency modules to the new EACF occurred during the winter, after a thorough testing of all systems and installations. However, the COVID-19 pandemic has restricted the activities of the OPERANTAR XXXIX in Antarctica, which focused on logistic activities, including the resupply and exchange of personnel of the EACF and the maintenance of the scientific equipment used by ongoing projects. The military personnel involved in the operation obeyed strict measures of quarantine and prophylaxis to avoid the introduction of contaminated personnel or material, which could provoke a crisis due to the lack of resources and the special conditions of the region.

Despite that all scientific activities in Antarctica were suspended in agreement with all the coordinators of scientific projects, some continued

[119] Decreto N° 9.759 (2019).

[120] Resolução N° 16 (2020).

remotely thanks to the collaboration offered by the Navy and some hydrographical surveys were undertaken as part of the regular operations of the *Maximiano*. Additionally, two new helicopters UH-17 (Airbus H135) of the Naval Aviation, replacements of the previous UH-13 (Airbus AS 355F2), were deployed to the *Maximiano*, making their maiden flight at the end of November 2020.

That renewal of Antarctic infrastructure in charge of the Navy—the new EACF, a new polar vessel to be constructed, and the replacement of the helicopters assigned to the Antarctic operations—will be supplemented by the modernization of the cargo aircraft through the replacement of the aged fleet of Hercules C-130 for the new KC-390, being developed by the Brazilian State aviation company Embraer. With all these investments, it is reasonable to expect that Brazil will increase its stakes on the white continent. However, after five years of cuts in science, a current policy of attack on the scientific sector and a tradition of imbalance between the priority given to the physical presence on the continent and the allegedly main activity in Antarctica, that is, scientific research, that expectation might be misplaced.

Only once the COVID-19 pandemic retreats and allows the scientific activity resumes, will it be possible to assess up to which point Brazil can maintain its commitment in Antarctica.

5.4 Conclusions

During the twenty-first century, favourable economic and political circumstances made the expansion of the PROANTAR and the consolidation of its scientific profile possible. Thanks to the impulse provided by some individuals in different institutions, the Brazilian Antarctic programme was able to seize the expansion of the national scientific system and the availability of resources. The strong environmental character that the Protocol of Environmental Protection gave to the Antarctic regime reverberated on the active involvement of the Ministry of Environment of Brazil in the Antarctic. At the same time, that participation opened the door to long-term planning and integrated projects, giving a more proactive character to the whole programme. This helped Brazil to secure a meaningful participation in the International Polar Year and to evolve towards a strategic planning of its Antarctic science.

However, the flexibility that the expansion and diversification of sources of finance gave to the programme was compensated for by the political dependence derived from the lack of autonomous guaranteed funding. That situation gave leverage to the programme allowing the relevant expansion of its activities and the renewal of its infrastructure, but it also made it increasingly dependent on the political will and the economic momentum. The maintenance of the image of Brazil's link with Antarctica through economic terms, either directly or mediated by environmental links, kept the instrumental role of science, and probably deprived the Antarctic science from justifying its value intrinsically. Furthermore, despite it has been some advancements in social sciences, the emphasis on the economy avoided a more thorough involvement in Antarctic humanities and social science research, which remained limited to the margins of the programme both politically and financially.

Also, the expansion of the programme to the interior of the continent was limited to the financial possibilities and revealed the pervasive maritime and coastal character that the Brazilian programme has today. The disparities between the political importance—and, consequently, financial resources—given to physical infrastructure and scientific research is also revealing of the remaining territorial character that the Brazilian programme has, and up to which point it is strongly linked in the public imagination to the availability of natural resources and the future possibility of its exploitation. At the same time, the resilience demonstrated by Brazil's presence in Antarctica after the accident at the EACF is an expression of a strong political will and the solid commitment of several groups within the scientific community, the Navy and different governmental circles. Notwithstanding, it also made apparent the strong contradiction that lay within the logic of the Brazilian Antarctic programme.

Despite the investment in the new EACF, the resources committed to the acquisition of a new vessel, and the intention to replace the aged fleet of Hercules C-130s by the new developed cargo aircraft being developed by Embraer, the KC-390, it is necessary to bear in mind that the main activity in Antarctica is scientific research, and that all logistic equipment is useless without its ultimate purpose of scientific research. The disparity between the resources applied to the acquisition of physical infrastructure and the resources allocated to scientific research made apparent the strong contradiction that lies within the Brazilian Antarctic programme and its limitations. The COVID-19 pandemic has delayed the tipping point when the resources for research would need to be renewed.

The fact that the PROANTAR has an important parliamentary support and it is led by the Navy could ensure that enough influence be exerted to obtain the necessary funds for logistic activities and research—which in comparison with the amount spent on infrastructure are minimal. However, presently, not even the cost of the operation of the logistics has been met in the last approved budget—that including Congressional amendments—and further funds will be needed to meet the financial needs of the operations.

Also, the cuts experienced in the scientific system affected the basis on which the entire Brazilian scientific and technological development depends.[121] Without grants to support post-graduate studies and professional prospects for those undertaking them, there is little incentive for people to start a scientific career or remain in the country. Even if the PROANTAR is able to find the financial resources for continuing its scientific programme, the continuity of a policy oriented to undermine the scientific system corrode the basis over which the programme develops and deprive it from the generation on which its future depends.

References

Abdenur, A. E., & Neto, D. M. (2014). Rising powers and Antarctica: Brazil's changing interests. *Polar Journal, 4*(1), 12–27. https://doi.org/10.1080/2154896X.2014.913910

Alvarez, C. E. de, Casagrande, B., & Soares, G. R. (2007). Resultados alcançados com a implementação do plano diretor da Estação Antártica Comandante Ferraz. In *IV Encontro Nacional e II Encontro Latino-americano sobre Edificações e Comunidades Sustentáveis* (pp. 1297–1306).

Becker, C., Cesar, C. M., Gallas, D., & Weber, M. H. (2016). Manifestações e votos sobre impeachment de Dilma Rousseff na primeira página de jornais brasileiros. *Revista Latinoamericana de Ciencias de La Comunicación, 13*(24), 96–113. https://www.alaic.org/revista/index.php/alaic/article/view/736

Cardone, I. J. (2014). A Política Antártica Brasileira no século XXI: evolução de fatores estruturais. In *Anais do Seminário Nacional Sociologia & Política*. UFPR.

[121] See, for example: A política brasileira de CT&I (2019).

At the time of writing this chapter, in January 2021, Bolsonaro has introduced a veto in the National Budget prohibiting the restraining and reallocation of the funds of the FNDCT on which the funding of many projects of science and technology, including a great part of the research funds of the PROANTAR, depends. Without that guarantee, the future of the Brazilian system of science and technology is severely compromised.

Cardone, I. J. (2015). *As Posições Brasileiras No Sistema Do Tratado Antártico Com Ênfase Na Questão Ambiental*. Universidade Federal de Parana. http://hdl.handle.net/1884/38844

De Almeida, A. T. S., & Lima, V. T. L. da S. (2016). Dilma Rousseff na Imprensa Brasileira: Da Reeleição ao Processo de Impeachment. *Encontros*, *14*(26), 102–113.

Ferreira, F. R. G. (2009). *O Sistema do Tratado da Antártica: evolução do regime e seu impacto na política externa brasileira*. Fundação Alexandre de Gusmão.

Krupnik, I., Allison, I., Bell, R., Cutler, P., Hik, D., López-Martínez, J., ... Summerhayes, C. (Eds.). (2011). *Understanding earth's polar challenges: International polar year 2007–2008*. CCI Press. https://doi.org/10.3723/ut.30.057

Marques, F. P. J., Mont'alverne, C., & Mitozo, I. B. (2018). A empresa jornalística como ator político: Um estudo quanti-qualitativo sobre o impeachment de Dilma Rousseff nos editoriais de Folha e Estadão. *Observatorio Journal*, *1*, 224–245. http://obs.obercom.pt

Mattos, L. F. de. (2014). A Inclusão da Antártida no Conceito de Entorno Estratégico Brasileiro. *Revista Da Escola de Guerra Naval*, *20*(1), 165–192. http://www.egn.mb/arquivos/revistaEgn/nova-revista/revista-20n1/index.htm

O Brasil e o Meio Ambiente Antártico / Ministério do Meio Ambiente. (2007). Ministério do Meio Ambiente.

Plano Mais Brasil: *Plano plurianual 2012–2015*. (2011). Ministério do Planejamento.

Rodrigues, T. M. (2018). The role of the media in the processes of impeachment of Dilma Rousseff (2016) and Michel Temer (2017). *Contracampo—Brazilian Journal of Communication*, *37*(2), 36–57. https://www.academia.edu/38001810/The_role_of_the_media_in_the_impeachment_processes_of_Dilma_Rousseff_2016_and_Michel_Temer_2017_

Scientific Committee on Antarctic Research Bulletin No. 161, December 2006. (2006). Cambridge.

Simões, J. (UFRGS), Viana, A. R. (CENPES), Secchi, E. R. (INPE/UPM), Correia, E. (INPE/UPM), da Silva, H. E. (UERJ), Wainer, I. E. K. C. (USP), ... Valentin, Y. Y. (UFRJ). (2013). *Ciência Antártica para o Brasil: plan de ação 2013–2022*. Brasília.

Van Dijk, T. A. (2017). How globo media manipulated the impeachment of Brazilian president Dilma Rousseff. *Discourse and Communication*, *11*(2), 199–229. https://doi.org/10.1177/1750481317691838

Other Sources

148ª Sessão Ordinária. (2002, April 23). [Proceedings]. Comissão Interministerial para os Recursos do Mar, Brazil.
149ª Sessão Ordinária. (2002, August 27). [Proceedings]. Comissão Interministerial para os Recursos do Mar, Brazil.
154ª Sessão Ordinária. (2003, December 2). [Proceedings]. Comissão Interministerial para os Recursos do Mar, Brazil.
Acordo de Cooperação Antártica entre o Governo da República Federativa do Brasil e o Governo da República do Chile. (2013, January 26). Brazil – Chile.
A Experiência da Engenharia da Marinha do Brasil na Parceria Científico-Tecnológica com a China nas Obras Civis para a Reconstrução da Estação Antártica Comandante Ferraz. (2018, September). DI/01 - XXIX RAPAL. RAPAL.
A política brasileira de CT&I e as manifestações da comunidade científica. (2019, December). Cadernos SBPC, Edição 2019. São Paulo: Sociedad Brasileira para o Progresso da Ciência.
Ata 157ª Sessão Ordinária. (2004, December 2). [Proceedings]. Comissão Interministerial para os Recursos do Mar, Brazil.
Ata 160ª Sessão Ordinária. (2005, December 1). [Proceedings]. Comissão Interministerial para os Recursos do Mar, Brazil.
Ata 161ª Sessão Ordinária. (2006, April 27). [Proceedings]. Comissão Interministerial para os Recursos do Mar, Brazil.
Ata 168ª Sessão Ordinária. (2008, May 13). [Proceedings]. Comissão Interministerial para os Recursos do Mar, Brazil.
Ata da 173ª Sessão Ordinária. (2009, December 1). [Proceedings]. Comissão Interministerial para os Recursos do Mar, Brazil.
Ata da 174ª Sessão Ordinária. (2010, April 27). [Proceedings]. Comissão Interministerial para os Recursos do Mar, Brazil.
Ata da 175ª Sessão Ordinária. (2010, August 26). [Proceedings]. Comissão Interministerial para os Recursos do Mar, Brazil.
Ata da 176ª Sessão Ordinária. (2010, November 24). [Proceedings]. Comissão Interministerial para os Recursos do Mar, Brazil.
Ata da 178ª Sessão Ordinária. (2011, August 23). [Proceedings]. Comissão Interministerial para os Recursos do Mar, Brazil.
Ata da 179ª Sessão Ordinária. (2011, November 29). [Proceedings]. Comissão Interministerial para os Recursos do Mar, Brazil.
Ata da 180ª Sessão Ordinária. (2012, April 24). [Proceedings]. Comissão Interministerial para os Recursos do Mar, Brazil.
Ata da 183ª Sessão Ordinária. (2013, April 23). [Proceedings]. Comissão Interministerial para os Recursos do Mar, Brazil.
Ata da 184ª Sessão Ordinária. (2013, August 29). [Proceedings]. Comissão Interministerial para os Recursos do Mar, Brazil.

Ata da 186ª Sessão Ordinária. (2014, April 24). [Proceedings]. Comissão Interministerial para os Recursos do Mar, Brazil.

Ata da 188ª Sessão Ordinária. (2014, November 25). [Proceedings]. Comissão Interministerial para os Recursos do Mar, Brazil.

Ata da 189ª Sessão Ordinária. (2015, April 23). [Proceedings]. Comissão Interministerial para os Recursos do Mar, Brazil.

Ata da 192ª Sessão Ordinária. (2016, April 29). [Proceedings]. Comissão Interministerial para os Recursos do Mar, Brazil.

Ata da 193ª Sessão Ordinária. (2016, August 30). [Proceedings]. Comissão Interministerial para os Recursos do Mar, Brazil.

Ata da 195ª Sessão Ordinária. (2017, April 26). [Proceedings]. Comissão Interministerial para os Recursos do Mar, Brazil.

Base da Antártida estava em festa quando incêndio começou em fevereiro. (2012, July 20). Uol Notícias [online]. https://noticias.uol.com.br/ultimas-noticias/agencia-estado/2012/07/20/base-da-antartida-estava-em-festa-quando-fogo-comecou.htm. Accessed on: 20 December 2020.

Chamada /Ação Transversal n°64. (2013, September 19). Programa Antártico Brasileiro. MCTI/CNPq/FNDCT. Brazil.

Chamada N° 21. (2018, August 22). Programa Antártico Brasileiro. CNPq/MCTIC/CAPES/FNDCT. Brazil.

Comitê Nacional de Pesquisas Antárticas é reativado. (2004, June 16). São Paulo, Agência FAPESP. https://agencia.fapesp.br/comite-nacional-de-pesquisas-antarticas-e-reativado/1967/. Accessed on: 10 December 2020.

Conceitos e Critérios Adotados para o Plano Diretor da Estação Antártica Comandante Ferraz. (2005, September). DI/51 - XVI RAPAL. RAPAL.

Considerações e Implicações da Adoção do Anexo VI (Responsabilidade) do Protocolo de Madri. (2005, September). DI/41 - XVI RAPAL. RAPAL.

Declaração Conjunta Brazil – Argentina. (2011, January 31). [Declaration signed in Buenos Aires]. Brazil. https://www.gov.br/mre/pt-br/canais_atendimento/imprensa/notas-a-imprensa/declaracao-conjunta-por-ocasiao-da-visita-da-presidenta-dilma-rousseff-a-republica-argentina-buenos-aires-31-de-janeiro-de-2011. Accessed on: 20 December 2020.

Decreto Legislativo N ° 597. (2009, August 28). Brazil.

Decreto N° 3.416. (2000, April 19). Brazil.

Decreto N° 5.484. (2005, June 30). Brazil.

Decreto N° 9.759. (2019, April 11). Brazil.

Defense White Paper. (2012). Brazil. https://www.gov.br/defesa/pt-br/arquivos/estado_e_defesa/livro_branco/lbdna_2013a_inga_net.pdf. Accessed on: 11 November 2020.

Diário da Cámara dos Deputados. (2003, December 11). ANO LVIII, N° 212. Brasília: Cámara dos Deputados, Brazil.

Ganhadores do Nobel enviam carta a Michel Temer com preocupações com cortes na ciência. (2017, October 2). G1 [online]. https://g1.globo.com/ciencia-e-saude/noticia/ganhadores-do-nobel-enviam-carta-a-michel-temer-com-preocupacoes-com-cortes-na-ciencia.ghtml. Accessed on: 27 January 2021.

Informe Final, XVIII RAPAL, Brasília, Brasil. (2007, September). XVIII RAPAL. RAPAL.

Informe Final. XXIII RAPAL. Rio de Janeiro, Brasil. (2012, September). XXIII RAPAL. RAPAL.

Iniciativas Ambientais Brasileiras. (2002). DI/21 - XIII RAPAL. RAPAL.

Janot: 'Não tenho dúvida de que Lula é corrupto. Dilma, não'. (2019, September 27). UOL Noticias [online]. https://noticias.uol.com.br/politica/ultimas-noticias/2019/09/27/janot-nao-tenho-duvida-de-que-lula-e-corrupto-dilma-nao.htm. Accessed on: 26 January 2021.

Lula parte para visita à base brasileira na Antártida. (2008, February 17). G1 [online]. http://g1.globo.com/Noticias/Politica/0,,MUL302245-5601,00-LULA+PARTE+PARA+VISITA+A+BASE+BRASILEIRA+NA+ANTARTIDA.html. Accessed on: 10 December 2020.

Lula volta da Antártica com a intenção de investir mais na base brasileira. (2008, February 18). G1 [online]. http://g1.globo.com/Noticias/Politica/0,,MUL303118-5601,00-LULA+VOLTA+DA+ANTARTICA+COM+A+INTENCAO+DE+INVESTIR+MAIS+NA+BASE+BRASILEIRA.html. Accessed on: 10 December 2020.

Marinha compra navio para base antártica. (2008, May 2). *O Estado de São Paulo* [online]. https://politica.estadao.com.br/noticias/geral,marinha-compra-navio-para-base-antartica,166248. Accessed on: 1 December 2020.

Medida Provisória N° 560. (2012, March 7). Brazil.

Navio Polar Almirante Maximiano. (n.d.). Acervo Arquivístico da Marinha do Brasil. http://www.arquivodamarinha.dphdm.mar.mil.br/index.php/navio-polar-almirante-maximiano. Accessed on: 10 January 2021.

O Censo de Vida Marinha Antártica: Iniciativa para o Ano Polar Internacional 2007/08 e Potencial de Colaboração entre os Países da América do Sul. (2005, September). DI/45 - XVI RAPAL. RAPAL.

Política Nacional de Defesa. (2012). Brazil. https://www.gov.br/defesa/pt-br/arquivos/2012/mes07/pnd.pdf. Accessed on: 11 November 2020.

Portaria Interministerial N° 1199. (2012, May 4). Ministry of Defense, Ministry of Science, Technology and Innovation and Ministry of Environment. Brazil.

Portaría N° 31. (2014, January 13). Ministério da Ciência, Tecnologia e Inovação. Brazil.

Programa Antártico Brasileiro. (2006, December 27). Portaria 318. Marinha do Brasil.

Proposta Brasileira de um Programa Latinoamericano para o Ano Polar Internacional. (2005, September). DI/39 – XVI RAPAL. RAPAL.

Relatório do Grupo de Trabalho para a proposição de um Planejamento Estratégico para o PROANTAR 2012–2022. (2011, November 29). Secretaria da Comissão Interministerial para os Recursos do Mar. Brazil.

Relatório Final: GT - Agenda Antártica. (2005). [PROANTAR five years plan] (Portaria MCT 397/2005). Brazil.

Resolução N° 2. (2010, August 26). Comissão Interministerial para os Recursos do Mar, Brazil.

Resolução N° 7. (2006, September 26). Comissão Interministerial para os Recursos do Mar, Brazil.

Resolução N° 16. (2020, July 30). Comissão Interministerial para os Recursos do Mar, Brazil.

Resultado da Avaliação Ambiental de Eficiência da Técnica Construtiva e do Planejamento Logístico no Desmonte do Refúgio Padre Balduíno Rambo. (2005, September). DI/53 - XVI RAPAL. RAPAL.

Se assumir, Temer promete reforma brusca, com mudanças em ministérios. (2016, April 14). Folha de São Paulo [online]. Source: https://www1.folha.uol.com.br/poder/2016/04/1760698-se-assumir-temer-promete-reforma-brusca-com-mudancas-em-ministerios.shtml. Accessed on: 26 January 2021.

Simões. (2018). Falta de recursos financeiros ameaça a continuidade da parte científica do Programa Antártico Brasileiro (PROANTAR) a partir de julho de 2018 – Análise de situação (2018) – Centro Polar e Climático/UFRGS. Source: http://www.diretodaciencia.com/wp-content/Anexos/Analise-Situacao-Proantar-2018.pdf. Accessed on: 7 January 2021.

South American Network on Antarctic Marine Biodiversity (BioMAntar). (2008, June). Information Paper submitted to the Committee for Environmental Protection (CEP) by Brazil - IP125/XXXI ATCM. Antarctic Treaty Secretariat.

Statement by Ambassador Luiz Alberto Figueiredo Machado. (2011). Final Report of the Thirty-fourth Antarctic Treaty Consultative Meeting, Buenos Aires 20 June-1 July 2011, pp. 576–577. Antarctic Treaty Secretariat.

Tribunal Militar condena sargento por incêndio em base da Antártica. (2016 May 16). G1 notícias [online]. Source: http://g1.globo.com/distrito-federal/noticia/2016/05/tribunal-militar-condena-sargento-por-incendio-em-base-da-antartica.html. Accessed on: 20 December 2020.

CHAPTER 6

Conclusions

6.1 Brazil's Trends in Its Antarctic Involvement

After examining the history of the almost seventy years since the Antarctic question in Brazil was first considered, and almost forty years since its active involvement on the continent, it is possible to identify some characteristics and trends that allow an interpretation of the changes and continuities of its Antarctic policy and to advance possible trends of its future evolution. Going through that history, the contradictions that have structured the way in which a country with strong tropical characteristics and identity have related to the Southern polar region are made apparent.

To begin with, in terms of the motives that pushed Brazil's involvement on the continent, one of the prevalent interpretations has been related to the geopolitical importance that the continent held in terms of national security and access to natural resources, especially strategic minerals. Whether it is undeniable that Brazil's interest in Antarctica was, and still is, framed by a perception of the geopolitical importance of the continent, it is its relatively low importance in comparative terms that explains the hesitations and limitations that the Brazilian approach to Antarctica shows over time. Therefore, the geopolitical interest of Brazil in Antarctica explains both its concerns and the relative marginal importance of the issue within its political priorities.

That limited geopolitical relevance is apparent when it is analysed the way in which factors related to Brazil's foreign policy—such as the

I. J. Cardone, *The Antarctic Politics of Brazil*,
https://doi.org/10.1007/978-3-030-80161-8_6

precaution to not antagonize Argentina and Chile—or its economic priorities—such as the need to come to an agreement in front of the conflict of the Itaipu dam—delayed the adoption of a positive agenda in Antarctica from the government. Furthermore, the government got to attempt against initiatives from the civil society—such as the efforts of the *Instituto Brasileiro de Estudos Antarticos* (Brazilian Institute for Antarctic Studies)—and some sectors of the intellectuality and the political elites—as the proposal elaborated by Teresinha de Castro and later supported by the Representative Euripides de Menezes.

Moreover, the official active involvement of Brazil in Antarctica followed the initiative of the Brazilian Navy and the projected vision of the future of the country that some of its elites envisioned and which was strongly related to the potential of the natural resources that could be seized from the seabed. The need for scientific knowledge to take advantage of such resources and enforce the control of their exploitation favoured the creation of the *Comissão Interministerial para os Recursos do Mar* (Inter-Ministerial Commission for Maritime Resources, CIRM) which adopted a scientific, interdisciplinary and inter-ministerial character.

However, at the same time, that relatively marginal relevance of the Antarctic in the geo-strategic scenario made it possible that, once the Cold War was over, Brazil remained involved in Antarctica and continued to pursue a scientific agenda which became its main strategic goal in the region. In other words, whether the concerns of national security fell with the Berlin Wall, and the economic prospects of obtaining gains from the exploitation of Antarctica's imagined 'fabulous' mineral resources disappeared after the Protocol of Environmental Protection to the Antarctic Treaty, Brazil's scientific interests in Antarctica came to prominence, becoming the ticket to the country's much desired development.

In this sense, some changes of the Antarctic attitudes were parallel to changes in the geopolitical visions projected from Brazil. Therefore, the formulations centred on National Security and geo-strategic concerns related to the bipolar competition between the Western and Communist blocs were replaced by the geo-economics perspectives, within which interest over the environmental questions gained increased relevance. Therefore, in the geopolitical realm, it is possible to identify an evolution from the idea of Antarctica as an issue of national security to its perception as a potential source of natural resources, finally arriving to the idea of environmental identity between Antarctica and Brazil.

In terms of Brazil's foreign policy, in contrast to the geopolitical thinking, Antarctica did not play any relevant role in the foreign policy agenda of Brazil up until well into the 1970s. Brazil's Ministry of Foreign Affairs (Itamaraty) ensured that no official position be taken and prevented the IBEA from embarrassing the Federal Government by taking autonomous action. In that sense, it is important to stress that the relative autonomy that Itamaraty held in the analysis and the role it played in steering the country's foreign policy guaranteed a cautious approach to the issue. At least until the last couple of years when that autonomy was put into question.

In general, Brazil's Antarctic policy went parallel with the main orientations of its foreign policy, despite it having maintained a certain degree of autonomy. It arose as part of the developmentalist model applied by the military regime, within which technological and scientific advancement was essential, and which envisioned a central role for the State. At the same time, it also reflected the position of Brazil and its intention to promote changes of the international system to make it more equitable, a claim that Brazil's brought to the Antarctic table, first with its claim about the need of a more open regime and afterwards with its demand that special conditions for developing countries should be considered in the mining agreement.

As the international system changed with the fall of the Soviet bloc, Brazil reoriented its foreign policy aiming for a deeper integration in the international system and acceptance of the international rules which emanated from the centre of power. This went in parallel with the abandonment of the developmentalist discourse within the Antarctic regime and the increasing search for regional coordination and cooperation. In addition, the spending cuts experienced during the 1990s as part of the neoliberal reform agenda impacted significantly Brazil's Antarctic programme, even when the character and relevance of the programme precluded it from being dismantled. Despite the calls for private companies' involvement, the impossibility of 'privatization' of the programme and the political costs that would imply to abandon it must have exerted an important influence in keeping the efforts that such involvement required.

When, entering the twenty-first century, Brazil rescued some of the principles that informed its developmentalist ideology—promoting public spending in social policy and technological and scientific development, and a more active role in the international realm—its Antarctic policy also was favoured with increased funding and widened political support.

With the progressive withdrawal of Brazil's international leadership during the second decade of the twenty-first century and cuts in public spending, Brazil's Antarctic programme suffered from financial instability that made the intervention of political and influential groups in order to guarantee its continuity, especially in what regarded its scientific activities, necessary. Recently, despite the recent departure from Brazil's trajectory on international politics and the adoption of a reactionary position in several international forums, Brazil's Antarctic programme has managed to hold its position thanks to the influence of the Navy and the political and influential groups that have backed up the programme over the last decades.

Contrasting with a relative low importance of the issue within the country's foreign policy priorities, its reluctance to give up an active role in the futures of the continent also reveals the place that Itamaraty coveted for the country in the international system and the relevance—even when limited—than conferred to Antarctica within it. In that regard, Brazil's foreign policy for Antarctica could be characterized as thoughtful and pragmatic, avoiding positions that could compromise its future interests and possibilities, leaving the options open when possible. This pragmatic approach proved to be very positive for the country, avoiding unnecessary political and economic costs but also granting a position at the table when considered relevant.

On the other hand, the intentional economy in the implementation of Brazil's Antarctic policy determined a relatively unspecialized institutional structure. Thus, the Brazilian policy for Antarctica made use of already existing structures, specially the CIRM, the Ministry of Foreign Affairs, the Ministry of Science and Technology, scientific agencies and funds and universities, among others. The specific institutional devices, such as the *Comisão Nacional para Assuntos Antárticos* (National Commission for Antarctic Issues, *CONANTAR*) and the *Comitê Nacional de Pesquisas Antárticas* (National Committee on Antarctic Research, *CONAPA*), have shown to be limited in terms of effectiveness and influence, having suffered from several disruptions in their work.

Within Itamaraty, the *Divisão do Mar, da Antártida e do Espaço* (Sea, Antarctic and Outer Space Division), has limited resources for a wide conglomerate of issues, with a high turn over of personnel. With some notable exceptions over time, there has been limited specialization of

Antarctic issues within Itamaraty's ranks, what explains the lack of the study of Antarctic politics in the country.

Also, the absence of an Antarctic institute or any centralized permanent structure has given the Brazilian programme great flexibility, allowing it to adapt to changing circumstances. Additionally, it has offered much autonomy to the scientific community for the definition of the scientific agenda. Nevertheless, the lack of a fixed structure promoting the Antarctic research and linking the scientific community and the political arena deprived the PROANTAR from a more constant influential force to lobby for resources and expand its horizons of operation. The creation of the National Institutes of Science and Technology of the Cryosphere and Antarctic Environmental Research was an important step in fixing specific scientific Antarctic structures. However, the inactivation of the latter revealed the limitations of the solution reached and called into question the need for some form of higher-level discussion between the different academic units involved in Antarctic research.

Probably the most revealing trait of the thin institutionalization of the Antarctic programme is the absence of a constituted memory of Brazil's policy and involvement in Antarctica. The geographical dispersion of the work developed in a diverse number of universities and research centres and the lack of a centralized permanent institution seems to have prevented the creation of special collections, photographic and multimedia archives, oral history records, documentary catalogues and even reference literature. This characteristic, as well as the relatively little emphasis given to humanities and social sciences, created a situation where the narratives about the Brazilian performance in Antarctica, its main protagonists, its evolution, its imaginaries and its achievements and limitations, among others, remains completely absent.

In terms of the alliances made, Brazil's Antarctic policy has proved very flexible, appealing to regional neighbourhoods, developing world identity or Antarctic active commitment, according to convenience and opportunities. A first phase could be characterized by a twofold approach, with the territorialist frontage theory advocates looking for a wide regional partnership—or a limited partnership that excluded Argentina and Chile if not possible—that gave Brazil regional leadership in Antarctica, and the official non-recognition of any claim to avoiding antagonizing Argentina and Chile. A second phase could be characterized by the search of a developing countries' alliance that could push Brazil's interest in the Antarctic oil exploitation potential, but without generating any strains with its regional neighbours and supporting the non-consultative Latin-American countries in their efforts to obtain full membership. A last, and still enduring, phase is that of the adoption of a Latin-American identity, strengthening regional coordination and cooperation and establishing institutional settings to that aim—nonetheless, without closing the door to other partnerships.

As regards the characteristics of Brazil's active involvement in Antarctica it is possible to highlight several aspects. The first and most outstanding is its maritime character. Despite the early designs of the policy that envisioned the more active involvement of the Army and the Air Force, the Army had been only marginally involved in the early stages, and the Air Force limited their actions to bring aerial support to the coastal operations of the Navy. Land operations have been entrusted to the *Clube Alpino Paulista* (São Paulo Alpinist Club) and no initiatives of acquiring capabilities—neither terrestrial nor aerial—to explore the interior of the continent have been registered during the research. The expeditions to the interior of the continent have been conducted thanks to the assistance of the Chilean programme and the initiative of Professor Jefferson Simões and the National Institute of Science and Technology of the Cryosphere.

Brazil's Antarctic programme emerged from an initiative of the Navy through the CIRM, which seems to have structured the programme in a lasting way. Despite the inter-ministerial and scientific character of the programme, the fact that it has been always fostered by the Navy has imprinted some characteristics that reveal the way in which Brazil's involvement in Antarctica has developed. That maritime character could be identified not only in the predominance of the Navy within the CIRM and its involvement in supporting the funding of the programme, but it also is revealed in other elements of the institutional design of Brazil for Antarctica. Most notable are the inclusion of Antarctica within the issues of Sea and Outer Space in a single department within Itamaraty, the grouping of Sea, Coastal region and Antarctica as a special item of the national budget, and the grouping within the twenty-first-century Strategic Environment of Brazil's Defence White Paper.

Another characteristic of Brazil's policy has been its focus on natural resources. Not only was Brazil's early thinking about Antarctica—the so-called frontage theory—closely related to the prospects of natural resource exploitation in Antarctica, but also its later involvement on the continent had been always linked with some form of economic interest, either direct or indirect. Early Brazilian involvement on the continent had much to do with the prospects of mineral exploitation—in particular oil—in view of the negotiations for an Antarctic mineral agreement. As the agreement did not satisfy Brazil's ambitions for acquiring special conditions that could balance the different levels of technological development, it turned rapidly to a position more favourable to a total ban of mineral activities. This took place at a time when its international image related to the protection of the environment was being questioned and it started to identify a more immediate economic interest in the ecological balance of the region due to its relationship with the agricultural activity in the national territory. At last, that economic interest was also linked to the scientific research developed in the region. In that sense, since the very beginning, the Brazilian programme has looked for research that could be linked to some economic interest for Brazil, becoming integrated in the later formulations of its scientific strategies in the form of better weather predictions, technological implementations and understanding of living natural resources dynamics.

Notwithstanding this maritime aspect of the Brazilian Antarctic programme and its interest in natural resources, its policy has been mainly scientifically driven, especially after 1991. This is not exclusive to Brazil as, despite the political and economic interests involved, Antarctic endeavour had always been related to the scientific enterprise.[1] But with the age of Antarctic exploration long over when Brazil first set its sights on Antarctica, the scientific character acquired more prominence. It was only once the country had developed a scientific system and acquired extensive capabilities that its Antarctic endeavour was made possible. This relative delay in developing a national scientific system can explain Brazil's later involvement. As well, its remarkable rapid development can also explain Brazil's success in becoming an active participant in the Treaty and a noteworthy scientific actor in Antarctica.

The beginning of Brazil's scientific efforts in Antarctica was mainly unplanned due to the swift nature of the initiative once the decision was taken. Projects were offered by invitation of the CIRM to well-regarded specialists in their respective disciplines, but who until then lacked Antarctic specialization or extensive polar experience. Notwithstanding, the rapid development of Antarctic expertise allowed the programme to

[1] Cardone (2019).

evolve in the following decades. The assumption by the National Research Council of the responsibility of the scientific aspects of the programme in 1991 and the creation of the *Comitê Nacional de Pesquisas Antárticas* (National Committee on Antarctic Research) in 1996 implied an explicit acknowledgement to the scientific community in their fields of expertise within the programme. But it would be only after the programme was integrated in issue-specific calls for scientific projects that the Brazilian Antarctic science acquired a stringer character. This evolved with the development of the two research networks, the planning of the Antarctic Agenda, the participation on the International Polar Year, the creation of the National Institutes of Science and Technology on Antarctic issues and the development of the scientific agenda 2013–2022.

On the other hand, the evolution of the scientific character and institutional development of the scientific structure of the programme were centred on the same main scientific fields that defined its development since the beginning with a tentative inclusion of alternative fields, in particular those related to humanities and social sciences. This could be attributed to the particular emphasis the programme gave to the economic aspect, which was much more indirect and harder to qualify in the case of the humanities and social sciences.

Another characteristic of Brazil's involvement in Antarctica has been the almost pervasive instability of funding. To some extent and linked to the economic instability of the national economy, it has also depended strongly on changes in the political arena and the lobby capabilities of some individuals within the programme. Whether the development ethos of the military regime ensured that the programme found resources in times of extreme instability and scarcity, the neoliberal reforms implemented with the return of democracy left the programme in a critical situation until a new impulse towards science and technology was assumed at the end of the twentieth century and the economic situation changed favourably. Still, the programme experienced various periods of expansion and retraction of funding, placing its continuity at stake on multiple occasions, even until today.

In this respect, the political alignment behind the programme has grown over time, widening its scope and providing more flexibility. As this allowed the programme to expand, it also exposed the programme to increased financial needs, which on several occasions have been met thanks to extraordinary funds obtained from the Congressional support group or by fund reallocations of the Navy's budget. This created a dependence that has forced the leaders of the programme to exert a continuous lobby in order to secure the necessary resources. This extended to logistic needs, scientific projects and political and technical participation on international events and forums.

However, a strong contradiction can be identified in the way in which Brazil's approach the investment in the physical presence in Antarctica and the way in which approach the investment on scientific activities. Brazilian perspective on physical presence in Antarctica seems to be attached to an outdated view—related to the geopolitical thought—by which the presence of physical infrastructure, such as an all-year station—is considered as necessary to remain an active member of the Antarctic regime. The same seems not to apply to the actual core activity required by the Antarctic regime to remain a consultative party, scientific research. That is not to say that infrastructure is not necessary to develop the scientific activity in Antarctica, as it surely does, but that Brazil's view of each element of its active presence is approached differentially in terms of financial needs. Whether the station and the polar ship obtained funding that exceeded USD 230 million, scientific projects have struggled with limited funding which amounted to USD 4.3 for a period of three years—extended to four years due to the COVID-19 pandemic.

With those limitations and the lack of a solid institutional setting, the Antarctic policy has been subject to the efforts dispensed by prominent individuals from several corners of Brazilian society and politics. Influential personalities within the Navy have been essential in the creation of the programme, but also to ensure its permanence and guarantee its continuity, even reallocating the internal—and sometimes also scarce—funds of the Navy for a programme that was not entirely its own. Politicians from various quarters have backed the programme and lobbied for resources, fostering the image of the programme as being strategic for the country. Also, well-regarded scientists have championed Brazil's Antarctic programme internationally, becoming Brazil's Antarctic banner due to the high level of development and influence reached in a short time.

The latter leads us to conclude that, whether Brazil's political participation in Antarctic politics has been rather reticent, especially after the 1990s, its scientific activity and physical presence in Antarctica have been noticeable worldwide, becoming a notorious Antarctic actor. It is true that, Brazil's presence in Antarctica has not been as intensified as it appeared originally, remaining largely modest, as argued in other work.[2]

Nonetheless, it is also true that today, Brazilian presence in Antarctica can be considered a consolidated programme, characterized by its high quality, environmental performance and solid trajectory of almost 40 years of uninterrupted presence. Whether some caveats remain about financial difficulties of the programme and the possible impacts that the current policy of attack to science could have, a look at the solid track record of resilience that the Brazilian programme suggests that it could weather any storm ahead.

6.2 Looking Ahead

The recent inaugurated Brazilian Antarctic station Comandante Ferraz, with its innovative design, cutting-edge technology, security and environmental systems and comfort, added to the call for bids for the construction of a new polar ship to replace the aged *Ary Rongel*, the replacement of the helicopters used for Antarctic operations for newer and more capable models, and the planned replacement of the Air transport fleet for the Embraer KC-390 seems to suggest a renewal of Brazil's interest in Antarctica. However, whether all these elements express the deep commitment of the country with Antarctica, they do not imply the acquisition of significant new possibilities in their Antarctic operations but the improvements of the existing capabilities in terms of safety, efficiency and comfort. In other words, rather than being an expression of the willingness to expand Brazil's involvement in Antarctica, these actions can be considered the consolidation of their current involvement.

The new EACF accommodates a similar number of personnel that the previous station and the emergency modules installed temporarily, but with much more comfort and better conditions for conducting their work. That includes laboratories equipped to produce some preliminary analysis and processes, and better installations to guarantee the conservation of samples, all in a much-improved infrastructure, designed with a view of a better stay for civil and military personnel.

[2] Sampaio et al. (2017).

The call for bids for the construction of the new polar ship includes specifications that will improve the efficiency and autonomy of sea operation, including more space for scientific personnel and activities, and a reduced crew, without compromising the safety of its operation. However, the requirements do not specify ice-breaking capabilities, rendering its theatre of operation relatively similar to that of the current ships.

The replacement of the helicopters and transport aircraft for newer models will improve efficiency but, in at first, will not result in any significant expansion of its operational area. The possibility of using the KC-390 to extend operations to the interior of the continent via ice-landing is subject to study, but with the plane still undergoing flight tests and final development, it will take some time—and the willingness of the Air Force in accepting the challenge—to ascertain up to which point that could be a real possibility.

On the other hand, that investment and modernization of Brazil's Antarctic infrastructure contrasts starkly with the disinvestment in science and the spending cuts planned by the government for the current year, throwing a shadow over the current prospects of Brazil's Antarctic programme and, also, its intermediate future. The suspension of the 2020–2021 season, Antarctic scientific activities *in loco* due to the COVID-19 pandemic have deferred the due date for the renewal of financial commitment to the scientific activities of the programme, and the failure experienced in meeting even the costs requiring for the resupply and maintenance of Brazilian infrastructure and basic operations in Antarctica certainly do not suggest a favourable panorama.

Worsening the situation, the cuts in science and technology in the last years have seriously compromised the social base in which an advanced scientific programme such as the PROANTAR depends. Would-be academic students lack incentives to pursue postgraduate studies and professionals from all disciplines start to join the ranks of the brain drain of Brazil in the face of the large reduction in the employment opportunities offered within the country, which adds to the climate of political and social persecution of sectors critical of the government.

With these conditions in mind, it is possible to foresee that Brazil's Antarctic programme will have to cope with some serious challenges in the years to come. The first challenge will be that of obtaining sufficient financial resources to complete the process of modernization of its infrastructure, cover the costs of its logistic operations and provide the necessary resources for the continuity of the Antarctic scientific activities.

It is very likely that the Brazilian Antarctic programme will eventually ensure the resources necessary for all the former, based on the influence of the Navy and other political sectors, and the consideration of the strategic importance of the programme. However, it is hardly the case that new projects, and especially a diversification of projects and research fields, could be put into operation in that context. The absence of a constituted field of Antarctic humanities and social sciences will certainly continue, as the field has been the most attacked sector within the Bolsonaro's government policy of confrontation with science, and no certain consensus about its need has yet been formed within Brazil's Antarctic community.

In the institutional arena, the solid performance of the National Institute of Science and Technology of the Cryosphere, the reorganization of the CONAPA and the role assumed by the CIRM in the face of the dissolution of the CONANATAR suggest that the traditional flexibility within which the Brazilian Antarctic programme rested still allow it to adapt to the shifting circumstances. However, it is hard to expect that substantial changes will be introduced in the near future.

The constitution of a working-group for the assessment of the 1987 POLANTAR could suggest that an important revision of Brazil's policy for Antarctica is underway. Certainly, the update of a pre-Protocol of Environmental Protection policy provides a unique opportunity to rethink the steering of Brazil's attitudes towards Antarctica. However, the lack of Antarctic politics specialists and in-depth studies on the issue, the restrictive participation given to interest groups and the relatively closed nature of that revision make it hard to expect a significant change in Brazil's orientations.

As regards its foreign policy, the current orientation of Brazil's suggests that no significant change will be induced in its Antarctic policy. Bolsonaro's government has been characterized by the rupture of Brazil's traditional positions in many international forums, but which have not reached Brazil's participation in the Antarctic regime. Despite the limitations that Itamaraty has shown to sustain its traditional independence on the steering of the foreign policy, the complexities involved and the lack of knowledge of the issue within Bolsonaro's ranks seems to have prevented the current government from giving new directions in Brazil's Antarctic policy.

In the short term, it is expected that Brazil's position and activities in Antarctica will remain mainly unchanged. However, in the long term, the current government's attitude to subverting Brazil's long-term international traditions and the erosion of the social base of its national scientific system can put its standing in Antarctica at risk. The evident subservient attitude of Itamaraty before the current government's designs for the country's foreign policy has not still threatened its position in the Antarctic international community, but suggests that it could suddenly change depending on random factors.

The active policy of disarranging the future generations of scientists and scholars by the government also poses a significant threat to the future of Antarctic science. Current project leaders and researchers still have a long road ahead, and the new capacities in terms of infrastructure suggest good possibilities for immediate development. However, if the current threat is not corrected and incentives are created in order to backtrack the current tendency of retraction of the area, it is very likely that once the current generation of academics near retirement, there will be no one to assume and continue their work.

In brief, until Brazil shows a more intense political will to raise its Antarctic profile, it is to be expected that it will continue to present the same limitations of the last decade. Moreover, the present orientations on foreign policy, science and public investment indicate important threats to the country's position in Antarctica, both politically and scientifically. Individual efforts to maintain an Antarctic 'lobby' at Congress and to sustain Antarctic investment will continue to be essential to the survival of the Brazilian Antarctic programme, but a correction of the current tendencies, especially in regard to the national policy on science and technology, is necessary for the future of the programme to be guaranteed. At last, for Brazil to close the parabola of connecting the tropic with the pole, it would need to assume a more assertive attitude in front of its Antarctic identity and trajectory, accepting the humanities and social sciences within the Antarctic specialities of worth and assuming the need to rescue its historical, cultural, social and environmental patrimony in the South Polar Region.

References

Cardone, I. J. (2019). *A continent for peace and science: Antarctic science and international politics from the 6th International Geographical Congress to the Antarctic Treaty (1895–1959)*. University of São Paulo—King's College London. https://doi.org/10.11606/T.101.2019.tde-16082019-164433.

Sampaio, D. P., Cardone, I. J., & Abdenur, A. E. (2017). Brazil, the Antarctic treaty system and Antarctica. In K. Dodds, A. D. Hemmings, & P. Roberts (Eds.), *Handbook on the politics of Antarctica*. Edward Elgar.

Index

A
Abdenur, Adriana Erthal, 17
Admiralty Bay, 103, 116, 118, 122, 176, 189, 207, 211
Admiralty Bay Antarctic Specially Managed Area (ASMA), 163, 164, 189
Africa, 16, 18, 205
Alaska, 149
Albuquerque, Lynaldo Cavalcanti (President of the CNPq), 96
Alcazar (Chilean vessel), 118
Almirante Câmara (Brazilian oceanographic vessel), 99, 145
Almirante Maximiano (Brazilian polar vessel), 202, 203, 213
  acquisition, 190
  adaptations, 193
  PROANTAR activities, 193, 206, 218
Amazon, 16, 18, 21
  blue Amazon, 94
  criticism over Brazil's environmental responsibility, 150, 159
American Antarctica, 36, 37, 43
Amorim, Celso (Brazilian diplomat), 206
Amundsen-Scott station (US), 107
Andersen, Rolf Trolle, 155
Andvord Bay, 116
Angola, Republic of, 60
Antarctic
  international conflicts, 35, 37
Antarctic Activities Support Station (ESANTAR), 115
Antarctic and Southern Ocean Coalition (ASOC), 149
Antarctica Project, 149
Antarctic economic resources, 54, 65, 87
  discussions at the SCAR, 87, 88
  marine living, 88
  mining, 55, 58, 61, 64, 88
  negotiations, 83, 88, 123, 143, 144, 148–149, 157
Antarctic environment
  concerns, 149, 152, 154
    environmental performance of Antarctic operations, 149

I. J. Cardone, *The Antarctic Politics of Brazil*,
https://doi.org/10.1007/978-3-030-80161-8

mining activities impact, 135, 136
ozone layer depletion, 135
conextions with, 51, 154
impacts, 149
Antarctic Peninsula, 35, 102, 111
Antarctic Studies Centre on Human Sciences, 193, 215
Antarctic territorial claims
Brazilian attitude in front of, 41, 51, 61, 63, 65, 140, 143
Antarctic tourism, 161, 178, 181, 185, 200
Antarctic Treaty, 197
accession, 58, 64, 85
articles, 44, 52, 58, 64
consultative status, 58, 109
negotiations, 41, 43, 44
revision, 61, 62, 64
Antarctic Treaty System, 180
Antarctic Treaty Consultative Meeting (ATCM), 88, 111, 124, 143, 150, 152, 158, 161, 180, 197, 209
Special Antarctic Treaty Consultative Meeting (SATCM), 85, 112, 119, 123, 143, 154, 157
Antarctic Treaty Secretariat, 161, 193
Convention on the Conservation of Antarctic Marine Living Resources (CCAMLR). *See* Convention on the Conservation of Antarctic Marine Living Resources
criticism over, 107, 108, 124, 135, 144
reactions of the Antarctic Treaty System, 111, 112, 136
opening of the regime, 144
Argentina, 14, 15, 21, 35–37, 40, 56, 60, 123, 138, 142, 196, 203, 231
Antarctic activities, 36
cooperation with Brazil, 109, 196, 206, 208
domestic conflicts, 57
incidents with Brazilian expedition, 103
Malvinas/Falklands War involvement, 99
misgivings about Brazil's intention, 92
negotiation of the Antarctic Treaty, 43
reaction to Poland's activities, 84, 85
Argentine-Chilean cooperation in Antarctic issues, 36
Artigas station (Uruguay), 123
Arturo Prat station (Chile), 103
Ary Rongel (Brazilian polar vessel), 184, 185, 191, 203, 204, 206, 216
acquisition, 161
adaptations, 162
ASK Subsea/Isis Viking (Russian company), 190
Atka Bay, 105
Australia, 84, 150
non ratification of CRAMRA, 148
Azambuja, Marcos Castrioto de (Brazilian diplomat), 112
Azambuja, Pericles (Brazilian journalist), 91, 92

B

Backheuser, Everaldo, 3, 5–6, 9, 14
Bahía Lasserre. *See* Admiralty Bay
Bahía Paraíso (Argentine polar vessel), 118, 149

Bahía San Blas (Argentine cargo vessel), 206
Bakker, Múcio Piragibe Ribeiro de (Brazilian Rear-Admiral), 95, 97, 101, 105, 110
Barão de Teffé (Antônio Luís Von Hoonholtz), 27, 101
Barão de Teffé (Brazilian polar vessel)
 acquisition, 99, 100
 adaptations, 110
 PROANTAR activities, 101, 102–106, 118–119, 119–120
 retirement from Antarctic activities, 161
Barão do Rio Branco, 28–30, 33
Beagle Channel, 103
Becker, Bertha, 8, 15–16
Belgrano station (Argentina), 40
Bellingshausen Sea, 48, 145
Bellingshausen station (USSR/Russia), 103, 118, 211
Berlin Colonial Conference (1884–1885), 39
Bernardo O'Higgins station (Chile), 211
Biological Investigations of Marine Antarctic Systems and Stocks programme (BIOMASS), 88, 100, 102, 110, 115, 119
Black Monday market crash, 143
Bolsonaro, Jair (President of Brazil), 215
 presidency, 215
  attacks to science, 215, 239
  effects over Brazil's Antarctic policy, 217
  foreign policy, 215, 238
  neo-liberal reform, 215
Borges, Durval Rosa (Brazilian reporter), 47
Bouvet Island, 91
Bransfield (British polar vessel), 94
Bransfield Strait, 100, 106, 115, 145, 204
Brasília, 163, 189
Brazil's concerns with the Antarctic environment, 113, 124, 152, 155, 159, 228
 effects over mainland economy, 167, 168, 180, 183, 208, 219, 233
 subordination to economic interests, 167
Brazil's economic interests in Antarctica, 55, 78, 101, 111, 126, 127, 140, 143, 160, 187, 199, 200, 229, 232
 living resources, 136, 137
 mining, 79, 97, 145, 156, 167, 231, 233
 tourism, 185
Brazil's environmental links with Antarctica, 67, 101, 113, 124, 127, 140, 152, 198, 206, 228
Brazil's environmental links with AntarcticaBrazil's environmental links with Antarctica, 208
Brazil's environmental policy
 effects over Antarctica, 151, 159
 international criticism, 150, 159, 233
Brazil's international agreements
 with Argentina, 196
 with Chile, 196, 208
 with the USSR, 142
Brazil's National Antarctic Programme (PROANTAR)
 advisory group, 164
 announcement, 93
 centrality of the Navy, 95, 96, 98, 126
 challenges, 237
 cooperation, 77, 101, 106, 196

environmental assessment group, 163, 165, 176, 179
institutional memory, 231
investment, 175, 179, 185, 188–192, 196, 214, 216, 218, 219, 229, 235, 236
financial difficulties, 160, 163, 167, 176, 177, 180, 184, 188, 195, 209, 213–215, 218, 220, 229, 230, 234, 235
management group, 106, 120
objectives and principles, 186
operational group, 120, 164
organization
acquisition of polar vessel, 97, 99, 156, 161, 190, 216, 237
political relevance, 178, 181, 198, 218, 219, 234, 236
problems of infrastructure, 175, 178
scientific programme, 77, 100, 127, 188
strategic planning, 181, 199, 218
working group PROANTAR, 194, 199–202
utilitarian character, 77, 82
Brazil's National Policy for Antarctica (POLANTAR)
accession to the Antarctic Treaty, 58, 61, 66
note of accession, 66
support from other parties, 62, 64
access to consultative status, 67, 75, 87, 112
background, 41, 42, 50
CCAMLR adoption, 137, 138
centrality of the Navy, 119, 159
contradictions, 167, 168, 219, 227, 235
cooperative scheme, 79, 231
design, 76
future prospects, 239
influence of frontage theory, 79, 82
institutional design, 75, 77, 80, 86, 98, 120, 126, 140, 141, 168, 230, 231
interests in Antarctica, 28, 39, 42, 46, 49, 51, 61, 62, 65, 67, 68, 78, 124
economic. *See* Brazil's economic interests in Antarctica
strategic. *See* Brazil's strategic interests in Antarctica
maritime approach, 97, 120, 126, 146, 198, 219, 232
participation of the scientific community, 119, 141, 159, 164
political principles, 77, 78, 80, 81, 186
pragmatic approach, 63, 80–82, 125, 140, 141, 167–169, 230, 231
reactions to the Antarctic Treaty, 42, 46, 49, 52, 61, 64, 66, 79
note of protest (1958), 42, 46, 49, 63, 65
regulation of Antarctic tourism, 178
reluctance to act, 43, 46, 50, 51, 55, 63, 68
reservation of rights, 140, 143
review, 139–141
support from other parties, 87
Working Group, 78, 85
Brazil's participation within the ATS, 113, 119, 124, 161, 168, 235
accession, 138
CCAMLR
accession, 136, 137, 168

developing country identity, 115, 124, 136, 139, 140, 143, 147, 158, 168, 229, 231
host of ATCM, 123, 143, 210
Brazil's strategic interests in Antarctica, 98, 100, 158, 160, 178, 198, 199, 228
strategic surrounding, 181, 205
Brazil's support of the Antarctic Treaty, 113–115, 139, 142, 159, 167, 197, 199
Brazil access to consultative status in the Antarctic Treaty, 107, 112
support from other parties, 109, 112
Brazilian Air Force, 80, 161, 184, 232, 237
Brazilian Antarctic Expedition (OPERANTAR)
Air Force operations, 111, 121, 122, 161, 194, 195, 211, 232
environmental management, 195
international exchange and cooperation, 101, 116, 203
OPERANTAR I, 101, 102–106
first cross of the Antarctic Polar Circle, 105
first enter into the Antarctic, 103
first official flight of the flag in the Antarctic, 103
incidents with other countries, 102, 104, 106
mechanical problems, 105
OPERANTAR II, 115–119
hydrographic survey, 118, 121
incidents with other countries, 118
OPERANTAR III, 121–122
hydrographic survey, 122
incidents with other countries, 122
OPERANTAR IV, 125
first wintering, 125
OPERANTAR V & VI
seismic surveys, 145
OPERANTAR XXIX, 195
OPERANTAR XXVIII, 193
OPERANTAR XXX, 202–204
incidents, 202, 204
OPERANTAR XXXI, 206–207
OPERANTAR XXXIII, 211
incident, 211
OPERANTAR XXXIX, 217
continuity of scientific work, 217
Brazilian Antarctic Institute (IANTAR), 76, 80, 97, 140
Brazilian Antarctic Science, 210, 234
autonomy, 160, 165, 191, 231
development, 161, 191, 192, 218, 233
financial difficulties, 160–162, 168, 177, 209, 213, 214, 220, 237
humanities and social sciences, 192, 200, 215, 219, 231, 238
instrumental role, 207, 219
international relevance, 75, 137, 141, 162, 179, 182, 214, 235
issue-specific research project, 176, 234
links with Brazilian territory, 65, 78, 81, 182, 187, 198, 200, 207, 209, 233
participation in the 4th IPY, 183, 184, 188
political relevance, 166, 168, 190, 199, 231, 233
promotion of the area, 106
research networks, 176, 184, 189, 234
strategic importance, 165, 228
Brazilian Army, 80, 232

Brazilian Constitution
  environmental protection, 151
Brazilian geopolitics
  development of geography, 4, 8–10, 18, 33
  development of geopolitical imagination, 10, 15–17, 18–21, 68, 228
  influence over policy, 1, 13, 15, 21, 34, 46, 59, 78, 167
  relative importance of Antarctica, 6, 8, 10–13, 14, 17, 18, 21, 37, 68, 227
Brazilian imperialism, 15, 56, 92
Brazilian Institute for Antarctic Studies (IBEA), 52, 56, 57, 64, 228, 229
  Antarctic expedition, 54, 61, 89
    joint project with Uruguay, 91
  Antarctic scientific activities, 57, 93
  declaration of public utility, 92
  Engineering Club of Rio de Janeiro, 52
  Rio Grande do Sul branch, 92, 93
  scientific programme, 54
  support in Parliament, 55
  view about Antarctica, 53
Brazilian Navy, 54, 80, 89, 121, 145, 177, 183, 184, 189, 191, 206, 228, 232, 234, 235
Brazilian presence in Antarctica, 47, 48
  Brazilian Antarctic Expedition (OPERANTAR). *See* Brazilian Antarctic Expedition (OPERANTAR)
  participation in other countries' expeditions, 93
  private initiatives, 52
Brazilian scientific system, 32, 34, 40, 68, 137, 233
  Brazilian Historical and Geographical Institute of Rio de Janeiro, 63
  cuts in investment, 237
  National Institutes of Science and Technology (INCT), 191, 192
  organization of scientists
    Brazilian Academy of Sciences, 141, 213
    Brazilian Society for the Advancement of Science, 35
    Brazilian Society for the Advancement of Sciences, 213
  participation of the Navy, 94
  scientific institutions, 35, 40
  universities, 184
    Fundação Universidade Federal do Rio Grande, 115
    Universidade Federal de Minas Gerais (UFMG), 193
    Universidade Federal do Rio Grande do Sul (FURGS), 180, 191, 196
    Universidade Federal do Rio Grande (FURG), 93
    Universidade Federal do Rio de Janeiro (UFRJ), 191
    Universidade de São Paulo, 32, 99, 100
    Universidade do Distrito Federal, 32
    Universidade do Vale do Rio dos Sinos (UNISINOS), 93, 125
Brazilian Superintendency of Fishing Development (SUDEPE), 137
Bremerhaven, 190
British Point (King George Island), 116

Brito, Tânia (Brazilian scientist), 176, 189
Brown station (Argentina), 103
Brussels, 209
Buenos Aires, 142, 193, 197
Bulgaria, 153

C

Cámpora, Héctor (President of Argentina), 57
Canberra, 112
Cardoso, Fernando Henrique (President of Brazil)
  Minister of Economy, 163
  Minister of Foreign Affairs, 164
  Presidency, 163, 167, 175, 176
  ratification of the Protocol of Environmental Protection, 165
Carvalho, Roberto de Guimarães (Brazilian Admiral), 184
Castro, Terezinha de, 10, 37, 43, 50, 53, 228
  frontage theory, 12, 44, 78
    influence, 79
Castro, Terezinha de, 12–13, 39–40
Chagasteles, Sergio Gitirana Florencio (Brazilian Commander), 166
Charcot, Jean-Baptiste, 27, 64
Chile, 14, 21, 35, 37, 123, 142, 184, 196, 203, 231
  Antarctic activities, 36, 49
  cooperation with Brazil, 49, 109, 196, 204, 208, 232
  negotiation of the Antarctic Treaty, 43
  offers of collaboration, 62
  reaction to Poland's activities, 84, 85
  role in the Protocol of Environmental Protection, 150
  Symposium internacional sobre el Desarrollo de la Antártica, 89, 94
Chilean Antarctic Institute (INACH), 196, 204
China Electronics Import and Export Corporation (Chinese company), 212, 216
China, Popular Republic of, 60, 168, 216
  accession to the Antarctic Treaty, 110
  transference of technology, 216
Christchurch, 106
Climate change, 200
Clube Alpino Paulista. *See* São Paulo Alpinist Club
Cold War, 10, 11, 13, 18, 34, 46, 64, 68, 158, 228
Collor de Mello, Fernando (President of Brazil), 155, 156
  impeachment, 160
Collor, Lindolfo L. (Brazilian Diplomat), 40
Comissão Interministerial para or Recursos do Mar (CIRM). *See* Inter-Ministerial Commission for Maritime Resources
Comissão Nacional para Assuntos Antárticos (CONANTAR). *See* National Commission for Antarctic Issues
Comité Nacional de Pesquisas Antárticas (CONAPA/CNPA). *See* National Committee on Antarctic Research
Congressional Group in Support of the Brazilian Antarctic Programme, 177, 178, 188, 189, 192, 234
Conselho Nacional de Desenvolvimento Científico e Tecnológico. *See*

National Research Council (CNPq)
Convention on the Conservation of Antarctic Marine Living Resources (CCAMLR), 136
Brazil's participation, 183
Brazilian attitude in front of, 168
conference, 88
Brazilian attitude in front of, 88
Convention on the Conservation of Antarctic Seals (CCAS), 139, 168
Brazil's ratification, 156
Convention on the Regulation of Antarctic Mineral Resources Activities (CRAMRA)
fall, 148–149
position of Brazil, 150, 152, 168
negotiations, 139, 143, 155
position of Brazil, 79, 123, 139, 147, 148, 151, 154, 168
signature, 147
Corpus dam, 56
Costa, Wanderley Messias, 15
Council of Managers of National Antarctic Programs (COMNAP), 153
background, 142
Cousteau, Jacques-Yves (French conservationist), 149
Covid-19 pandemic, 217–219, 237
Crisis of the 'Asiatic Tigers', 167
Cryosphere expedition, 196, 204
Cunha Bueno, Antônio Sílvio (Brazilian Federal Representative), 49
Cunha, Eduardo (Brazilian politician), 212
Curitiba, 182
Czechoslovakia, 153
Czech Republic, 211

D
da Costa, João Frank (Brazilian diplomat), 51
Dakshin Gangotri station (India), 108
Deception Island, 57, 106
Defence White Paper, 205, 232
Delgado de Carvalho, Carlos, 3, 4, 9, 38–40
Developing countries, 79
Divisão do Mar, da Antártica e do Espaço (DMAE). *See* Itamaraty: Seas, Antarctic and Space Division
Dorian Bay, 116
Doumer Island, 116
Drake Passage, 44, 51, 103, 156, 204

E
Ecuador, 123, 142, 164
Eduardo Frei station (Chile), 103, 204
Eisenhower, Dwight D. (US President), 42
Elephant Island, 57, 118, 121
Ellsworth Station (US), 40
Eltanin (US polar vessel), 55, 96
Embraer KC-390, 218, 219, 236, 237
Emerging power, 13, 16
Endurance (British polar vessel), 36, 94, 99, 118
negotiation for acquisition, 96
Engenavi (Brazilian company), 156
Engineering Club of Rio de Janeiro. *See* Brazilian Institute for Antarctic Studies
Environmental activist groups, 135, 150
Environmental Defence Fund, 149
Ernest Shackleton (British polar vessel), 204

Escola Superior de Guerra (ESG). See War College
Esquadrão Gordo. See Brazilian Antarctic Expedition (OPERANTAR)
Estação Antártica Comandante Ferraz (EACF), 202
  adaptations, 121, 122, 125, 156, 194
  experimental ethanol generator, 202
  fire at the station, 202, 203
  installation, 102, 105, 113–114, 117
    location, 116–117, 126
  installation of emergency modules, 207
  reconstruction, 204
  removal, 207, 210
Estação Antártica Comandante Ferraz (EACF), 190
  adaptations, 180, 183, 184
  communication system, 180, 183, 184
  installation
    location, 110
  master plan, 175, 176, 183
  new station, 236
  problems, 175, 183
  reconstruction, 208, 213, 216
    bidding, 211–212
    transference to new station, 217
  removal, 203
Estação de Apoio Antártico (ESANTAR). *See* Antarctic Activities Support Station
Estado Maior das Forças Armadas (EMFA). *See* Joint Chiefs of Staff
Estúdio 41 Arquitectura SS Ltda (Brazilian company), 208
Exxon Valdez (oil transport vessel), 149

F
Falcão, Armando (Brazilian politician), 66, 75
Falkland Islands Dependencies Survey, 37
Fanta, Edith (Brazilian scientist)
  chair of the Scientific Committee of the CCAMLR, 182
  member of the Joint Committee of the 4th IPY, 182
  SCAR Representative to the CCAMLR, 182
Faraday station (UK), 103
Felinto Perry (Brazilian submarine rescue ship), 195, 206
Ferraz (Brazilian Antarctic Station). *See Estação Antártica Comandante Ferraz (EACF)*
Ferraz, Luiz Antônio (Brazilian Naval Commander), 94, 100, 116
Ferreira, Felipe Rodrigues Gomes (Brazilian diplomat), 166
Figueiredo, João Batista (President of Brazil)
  background, 51
  presidency
    relationship with Argentina, 89, 90
  role in the PROANTAR, 101, 116
Fonseca, Maximiano Eduardo da Silva (Brazilian Admiral), 94
  resignation, 119
  role in the PROANTAR, 101, 114, 116, 126
France, 150
  non ratification of CRAMRA, 148
Franco, Itamar (President of Brazil)
  Presidency, 160
  vice-Presidency, 160
Frente Parlamentar de Apoio ao Programa Antártico Brasileiro. See Congressional Group in

Support of the Brazilian Antarctic Programme
Frontage theory, 12, 14, 39, 41, 46, 50, 51, 53, 65, 67, 82, 110, 125, 158, 231
foreign repercussions, 53
rebuttal, 91, 111, 141
Frontage theory Frontage theory, 44
Fundação de Estudos do Mar (Brazilian institution), 185

G
Geisel, Ernesto Beckmann (President of Brazil), 58
accession to the Antarctic Treaty, 66, 68
Apertura (opening), 59
presidency, 60
president of Petrobras, 60
Geopolitics, 5
and Antarctica, 1
Gerlache, Adrien de, 27
fly of the Brazilian flag, 63, 64
Gerlache Strait, 102, 103, 116
Germania (Merchant vessel), 206, 207
Germany
Democratic Republic, 143
Revolution, 153
fall of the Berlin Wall, 153
Federal Republic of, 60, 84, 123
membership to the SCAR, 87
under Nazi rule, 33, 37
Getúlio Vargas, 6, 32–34
Glasnost, 142
Glomar Challenger (US research vessel), 58, 65
Gorbachev, Mikhail (Russian politician), 142
Goulart, Jõao (President of Brazil), 50
Great Wall station (China), 123, 211
Greenpeace, 150, 156
World Park Base, 149
Greenwich Island, 103
Growth Acceleration Programme, 195
Guerreiro, Antonio (Brazilian diplomat), 147
Gulmar Atlantis (Brazilian submarine support vessel), 203

H
Haiti, 194
Halley Station (UK), 40
Hawke, Bob (Australian Prime Minister), 148
Henryk Arctowski station (Poland), 85, 103, 106, 116, 118, 119
Hero (US polar vessel), 57, 118
History of Brazil
conservatism, 50
economy, 6, 32, 33, 58, 95, 138, 143, 155, 161, 163, 167, 179, 195
foreign policy, 30, 33, 59, 69, 89, 156
relationship with Argentina, 90
responsible pragmatism, 59
territorial sea, 94, 95
foreign policy, 28–30
general history, 3, 7, 123, 195
coronelismo, 3, 7, 31
1930 Revolution, 6, 32
1964 military coup, 50
Collor's impeachment, 160
return to democracy, 158
Rousseff's impeachment, 212
general history, 31–32
military regime, 58, 59, 78
of modernism/positivism, 4, 7, 8, 31, 34
Hobart, 137
Hungary, 153

I
IBAS, 181, 193
India, 120, 168, 181, 193
accession to the Antarctic Treaty, 112
Antarctic activities, 108
CCAMLR, 138
contrast with Brazilian attitudes towards Antarctica, 108
parallelisms with Brazilian attitudes towards Antarctica, 107
proposal of the Antarctic Question at the UNGA, 40, 107
Instituto Antártico Brasileiro (IANTAR). *See* Brazilian Antarctic Institute
Instituto de Arquitetos do Brasil, 208
Instituto Nacional de Ciencia e Tecnologia Antártico de Pesquisas Ambientais. *See* National Institute of Science and Technology for Antarctic Environmental Research
Instituto Nacional de Ciencia e Tecnologia da Criósfera. *See* National Institute of Science and Technology of the Cryosphere
Inter-American Treaty of Reciprocal Assistance (TIAR), 37, 43, 46, 67, 78
Inter-Ministerial Commission for Maritime Resources (CIRM), 98, 140, 146, 165, 185, 186, 199, 228, 230, 232
competences and responsibilities, 98, 159, 160, 179
history, 95
role in the PROANTAR, 166, 184, 238
Secretariat (SECIRM), 95, 114, 184, 194
role in the PROANTAR, 166
subcommittee of the PROANTAR, 97, 165, 183, 185
working group POLANTAR, 217, 238
International Exhibition of Polar Exploration, 35
International Geophysical Year (IGY), 38–41, 43
International Institute for Environment and Development, 149
International Monetary Fund, 167
International Polar Year
First, 101
Fourth, 179–181, 185, 188, 218, 234
Brazil's participation, 181, 183, 184, 188
International Geophysical Year. *See* International Geophysical Year
Isla 25 de Mayo. *See* King George Island
Islas Orcadas (Argentine polar vessel) *see alsoEltanin*, 96
Itaipu dam
dispute with Argentina, 52, 56, 60, 69, 89
effects over Brazil's Antarctic project, 56, 60, 83, 126
negotiations, 83
understanding with Paraguay, 56
Italy, 33, 143
Itamaraty, 28, 30, 33, 40, 49, 51, 76, 80, 81, 87, 89, 97, 120, 165, 166, 184, 185, 194, 215, 230
accession to the Antarctic Treaty, 61, 63–65
autonomy, 69, 229, 239
reactions to the IBEA's project, 61
report to the Presidency, 62

Seas, Antarctic and Space Division (DMAE), 120, 146, 230, 232
Itamaraty, 28

J
Japan, 88, 146
Joint Chiefs of Staff (EMFA), 50, 51, 60, 76, 81, 90, 95
Jubany/Carlini station (Argentina), 106, 118, 211

K
Kacowicz, Arie, 1, 2, 21
Karol, Eduardo, 16
Keller Peninsula (King George Island), 116
King George Island, 57, 85, 103, 116, 125, 176
King Sejong station (South Korea), 211
Kirchner, Cristina Fernandez de (President of Argentina), 196
Klink, Amyr (Brazilian explorer), 178
Korea, Republic of, 138
Krill, 84, 88, 136
Kyoto, 161

L
Lastiri, Raúl Alberto (provisional president of Argentina), 57
Latin-American cooperation, 29, 30, 44, 168, 232
  in Antarctica, 65, 141, 189, 192, 196, 205, 229
Laurie Island, 36
League of Nations, 30, 34
Leal, Jorge Edgar (Argentine General), 92
Lima, 163
Livingston Island, 57, 206
Lula da Silva, Luiz Inácio (President of Brazil)
  Presidency, 16, 179, 188, 195
    Antarctic policy, 177, 190

M
Machado, Anésia Pinheiro (Brazilian female pilot), 47
Machado, Luiz Alberto Figueiredo (Brazilian diplomat), 197
Madrid, 157
Malaysia
  initiative to replace the Antarctic Treaty regime, 108, 112, 149, 152
Malvinas/Falklands War, 99
  aftermath, 102, 122
Mar de la Flota. *See* Bransfield Strait
Mar sem Fim (Brazilian tourist motorboat), 204, 206
Marsh station (Chile), 111, 117, 118
Mattos, Carlos de Meira, 10, 13–15
Maury, Matthew Fontaine, 27
Maxwell Bay, 204
McMurdo station (US), 106
Meeting of Administrators of Latin American Antarctic Programmes (RAPAL), 153, 158, 168, 211
  background, 141
  Brazil's participation, 163, 166, 183–184
    host of meeting, 163, 166, 189, 205
Meeting of the heads of the National Antarctic Institutes of Argentina, Chile and Uruguay. *See* Meeting of Administrators of Latin American Programmes
Menezes, Euripides Cardoso de (Brazilian Federal Representative), 50, 228

Mercado Común del Sur. *See* South American Common Market
Ministry of Agriculture, 81
Ministry of Communications, 76
Ministry of Defence, 198, 213
Ministry of Education and Culture, 185, 194
Ministry of Environment, 167, 175, 176, 184, 189, 192, 194, 195, 198, 204, 207, 218
  involvement in the PROANTAR, 176
  National Department of Environment, 76
Ministry of Foreign Affairs. *See* Itamaraty
Ministry of Justice, 69
Ministry of Mines and Energy, 76, 81, 120
Ministry of Science and Technology, 137, 141, 164, 166, 184, 191, 194, 198, 209, 213, 230
Ministry of the Air, 76
Ministry of the Army, 76
Ministry of the Navy, 76
Ministry of Tourism, 185, 194
Miyamoto, Shiguenoli, 17
Montevideo, 139, 142, 163
Mota Jr, David Silveira da (Brazilian diplomat), 124
Mouron, Milton (Brazilian Vice-President), 216
Musso, Julio (Uruguayan academic), 91
  misgivings from the Government, 91

N

National Commission for Antarctic Issues (CONANTAR), 80, 98, 140, 159, 185, 186, 230
  dissolution, 217, 238
  meetings, 100, 124, 185, 210
    suspension, 165, 166, 192
  regimentation, 109, 137, 141
National Committee on Antarctic Research (CONAPA), 186, 230, 234
  background, 119
  dissolution, 217
  institution, 164, 165
  meetings, 179, 185
    suspension, 165, 166
  reactivation, 217, 238
  regimentation, 181, 210
  strategic planning, 207, 211
National Defence Policy, 16, 17, 165, 181, 205
National Department of Meteorology, 76
National Fund for Scientific and Technological Development, 209
National Institute of Science and Technology for Antarctic Environmental Research, 191, 195, 231
National Institute of Science and Technology of the Cryosphere, 191, 196, 214, 231, 232, 238
National Pluriannual Budget, 198
National Research Council (CNPq), 35, 51, 77, 80, 81, 89, 106, 119, 159–161, 164–166, 184, 189, 192–194, 209, 234
National Security, 228
National Security Council, 62, 63, 66, 69, 76, 80–82, 92, 97, 139
National Security Doctrine, 21, 34, 46, 139
Negotiations on Antarctic mining. *See* Convention on the Regulation of Antarctic Resources Activities
Neiva, Eugênio Ferreira (Brazilian Lieutenant Commander), 97

Neto, Danilo Marcondes, 17
Neto, Júlio Soares de Moura (Brazilian Commander), 190, 205
Neumayer station (Federal Republic of Germany), 102, 105, 114
Neves, Tancredo (Brazilian politician), 123
New Zealand, 107
9th Biology Symposium, 182
1989 Revolutions, 153
Non-Aligned Movement, 108, 115, 149, 152
Nordenskjöld, Otto, 36
Norvegia Cape, 91
Norway, 67, 146

O

Ocean Empress. *See Almirante Maximiano*
Oi (Brazilian company), 192
Operação Antártica Brasileira (OPERANTAR). *See* Brazilian Antarctic Expedition
Operation Tabarin. *See* United Kingdom

P

Palmer Archipelago, 116
Palmer station (US), 103
Paris, 150, 152
Peixoto, Ernani do Amaral (Brazilian diplomat), 43
Perestroika, 142
Peron, Juan Domingo (President of Argentina), 56
Peru, 123, 142, 164
Petrobras, 58, 60, 145, 192
Piloto Pardo (Chilean vessel), 118
Pinheiro, João Batista (Brazilian diplomat), 61
Pinto Coelho, Aristides (Brazilian scientist), 52, 53, 57
Poland, 88, 123, 176, 203
  accession to the Antarctic Treaty consultative status, 84, 85
  Antarctic activities, 84
  effects on Brazil's candidacy, 85
  membership to the SCAR, 87
  offers of collaboration, 116
  Revolution, 153
Polar Circle (Norwegian polar vessel), 108
Polar Queen (Norwegian polar vessel). *See Ary Rongel*
Política Nacional para Assuntos Antárticos (POLANTAR). *See* Brazil's National Policy for Antarctica
Port Lockroy, 116
Princess Martha Land, 105
Prof. W. Besnard (Brazilian oceanographic vessel), 99, 100, 191
  adaptations, 110
  PROANTAR activities, 115, 118
Prof. W. Besnard (Brazilian oceanographic vessel), 102, 155
  PROANTAR activities, 106
Programa Antártico Brasileiro (PROANTAR). *See* Brazil's National Antarctic Programme
Protocol of Environmental Protection to the Antarctic Treaty, 158, 168, 195, 207
  Annex on Responsibility, 161, 180
  Brazil's attitudes, 163, 164, 166, 200
  Committee of Environmental Protection, 189
  effects over Brazil's policy, 175, 179, 186, 218
  negotiations, 154, 157

origins, 149, 152, 154
signature, 158
Punta Arenas, 102, 103, 106
Py Point, 116

Q

Quadros, Jânio (President of Brazil), 50
Quintal, Josias (Brazilian politician), 177
Quito, 163

R

Reunión de Administradores de Programas Antárticos Latinoamericanos (RAPAL). *See* Meeting of Administrators of Latin American Programmes
Ribeiro, Professor Joaquim, 37
Ricúpero, Rubens (Brazilian diplomat), 96
Rio de Janeiro, 101, 106, 119, 122, 205, 207
Rio Grande (Brazilian port), 102, 106, 122
Riso Patrón, Luis, 43
Rocard, Michel (French President), 148
Rocha-Campos, Antonio (Brazilian scientist), 106, 194
president of the SCAR, 162
secretary of the SCAR, 137, 141
Romania, 153
Rousseff, Dilma (President of Brazil)
foreign policy, 196
impeachment, 212
Presidency, 195

S

Santiago de Chile, 142
Santos (Brazilian port), 102, 115
São Paulo Alpinist Club, 101, 232
Saraiva Guerreiro, Ramiro Elysio (Brazilian diplomat), 100, 101, 111, 116, 136
Sarney, José (President of Brazil), 123, 142
Scientific Committee on Antarctic Research (SCAR), 83, 137, 141, 142, 153, 176, 179, 181, 207
membership, 76, 80, 119, 165
observer, 87
Scott Base (New Zealand), 106
Secretaria da Comissão Interministerial para or Recursos do Mar (SECIRM). *See* Inter-Ministerial Commission for Maritime Resources
Secretary of Aquiculture and Fishing, 210
Secretary of Environment, 159
Secretary of Planning of the Presidency, 76, 81
Setubal, Olavo Egydio (Brazilian politician), 124
Shackleton Base (UK), 40
Shackleton, Sir Ernest, 36
Shapley, Deborah, 61
Shelter astrônomo Cruls (Brazil), 122
Shelter engenheiro Wiltgen (Brazil), 122
Shelter Padre Balduíno Rambo (Brazil), 125, 156, 178
removal, 180
Silva, Golbery de Couto e, 10–12
Chief of Staff of Geisel, 59
hemicycles, 11
Silva, Luiz Inácio (President of Brazil). *See* Lula da Silva, Luiz Inácio
Silveira, Antonio Azeredo da (Brazilian diplomat), 76, 90

Simões, Jefferson Cardia (Brazilian scientist), 106, 180, 196, 204
vice-president of the SCAR, 214
Sociedade Anonima de Equipamentos Ferroviários (SAEF), 114
Sodré, Roberto de Abreu (Brazilian politician), 143
South Africa, 181, 193
South American Common Market (MERCOSUR), 158, 163
South Atlantic, 187, 199
Southern Ocean, 187, 199
South Orkneys, 36
South Pole, 48, 92, 107, 180
South Shetlands, 57, 100
South-South cooperation, 14, 16
Spain, 211
Special Secretary of Aquaculture and Fishing, 182
Station G (UK), 116, 122, 164
Stinker Point (Elephant Island), 121
Stockholm, 180

T
Telemar (Brazilian company), 184, 192
Temer, Michel (President of Brazil)
presidency
cuts on science and technology, 213
neo-liberal reform, 213
role in Rousseff impeachment, 212
Teniente Camara station (Argentina), 206, 208, 211
Teniente Luis Carvajal Villaroel station (Chile), 196
Teoria da defrontação. *See* Castro, Terezinha de
Thala Dan (Danish polar vessel). *See* Barão de Teffé (Brazilian polar vessel)
Thatcher, Margaret (British Prime Minister), 96
Travassos, Mario, 6–8
Aspectos Geográficos Sul-americanos, 6
Tripartite declaration (Argentina, Chile, UK), 37
Tropic, 18
geographical characteristics, 28
links with Antarctic region, 187, 206, 227
tropical imagination, 28
Trump, Donald (US President), 216

U
Union of Soviet Socialist Republics, 41, 88
United Kingdom, 35, 40, 203, 211
Antarctic activities, 37
incidents with Brazilian expedition, 102, 106, 118, 122
Malvinas/Falklands War involvement, 99
offers of collaboration, 67, 110
Operation Tabarin, 37
United Nations Convention on the Law of the Sea (UNCLOS), 95, 145
United Nations (UN), 33
Conference on Environment and Development, 152, 155, 159
India's Proposal on the 'Antarctic question', 40, 107
Malaysia's proposal on the 'Antarctic question', 108, 110, 112, 149, 152
'question of Antarctica', 114
United States, 29–30, 33, 37, 40, 42, 50, 146, 157
Antarctic activities, 47–49
National Science Foundation, 55
offers of collaboration, 55, 106

Uruguay, 123, 142, 168, 203
Antarctic activities, 122
private initiatives for joint expedition with Brazil, 91
Uruguayan Antarctic Institute (IAU), 91
Ushuaia, 106
USSR, 123, 146
cooperation with Brazil, 142
crisis, 142, 153
limited democratization, 142

V
Vale Soluções em Energia (Brazilian company), 202
Valle, Henrique (Brazilian diplomat), 152, 154, 159
Vasconcelos Torres, João Batista de (Brazilian Senator), 55
Vaterloo Island. *See* King George Island
Venice, 161
Villela, Rubens Junqueira (Brazilian scientist), 48, 53, 54
Viña del Mar, 152, 154

W
Wall Street stock market crash, 32
War College (Brazil-ESG), 10, 12, 13, 15, 34, 41, 52
Washington Conference (1959). *See* Antarctic Treaty
Washington Consensus, 155, 156, 158
Weddell Sea, 40, 102, 105, 111
Wellington, 143
Western Geophysical (US Company), 146
Westland Wasp (helicopters), 102
Wiencke Island, 116
Wilderness reserve, 149
Wiltgen, João Aristides (President of the IBEA), 52, 91, 92
Workers' Party, 212
World Bank, 150
World Park Base. *See* Greenpeace
World War I, 30, 31
World War II, 11, 18, 32, 33, 35, 37

Z
Zarankin, Andrés (Argentine scientist), 193

The manufacturer's authorised representative in the EU is Springer Nature Customer Service Centre GmbH, Europaplatz 3, 69115 Heidelberg, Germany. If you have any concerns regarding our products, please contact ProductSafety@springernature.com

Printed and bound by CPI Group (UK) Ltd, Croydon, CR0 4YY
30/09/2026
02225701-0007